GEOMETRY
Theorems and Constructions

GEOMETRY
Theorems and Constructions

ALLAN BERELE

JERRY GOLDMAN

DePaul University, Chicago

PRENTICE HALL, *Upper Saddle River, New Jersey 07458*

Library of Congress Cataloging-in-Publication Data

Berele, Allan.
    Geometry: theorems and constructions / Allan Berele, Jerry Goldman.
        p. cm.
    Includes index.
    ISBN 0-13-087121-4
    1.   Geometry. I. Goldman, Jerry, 1939– II. Title.
    QA453.B518 2001
    516-dc21   00-060653

Acquisition Editor: *George Lobell*
Editor-in-Chief: *Sally Yagan*
Production Editor: *Bayani Mendoza de Leon*
Assistant Vice President of Production and Manufacturing: *David W. Riccardi*
Executive Managing Editor: *Kathleen Schiaparelli*
Senior Managing Editor: *Linda Mihatov Behrens*
Manufacturing Buyer: *Alan Fischer*
Manufacturing Manager: *Trudy Pisciotti*
Marketing Manager: *Angela Battle*
Marketing Assistant: *Vince Jansen*
Director of Marketing: *John Tweeddale*
Editorial Assistant: *Gale Epps*
Art Director: *Jayne Conte*
Cover Designer: *Bruce Kenselaar*
Cover Image: *Strode Eckert/Zimmer Gunsul Frasca Partnership*

© 2001 by Prentice Hall, Inc.
Upper Saddle River, New Jersey 07458

Printed in the United States of America
10 9 8 7 6 5 4 3 2 1

ISBN: 0-13-087121-4

Prentice-Hall International (UK) Limited, *London*
Prentice-Hall of Australia Pty. Limited, *Sydney*
Prentice-Hall of Canada Inc., *Toronto*
Prentice-Hall Hispanoamericana, S.A., *Mexico*
Prentice-Hall of India Private Limited, *New Delhi*
Prentice-Hall of Japan, Inc., *Tokyo*
Pearson Education Asia Pte. Ltd.
Editora Prentice-Hall do Brasil, *Ltda., Rio de Janeiro*

Contents

Preface

Our main goals in writing this textbook were to teach geometry, to teach about proofs, and to teach beautiful mathematics. And, of course, we wanted to write a book that was honest, which is to say rigorous without being pedantic. As obvious as these goals seemed to us, we didn't know of another book that did things this way.

Many recent college geometry books have been centered on foundations of geometry, non-Euclidean geometry, convex sets, transformation methods, and the like. Valuable as these topics are, the end result is that these texts give Euclidean geometry short shrift. Secondary school teachers are going to be teaching Euclidean geometry, they need to know it, and it is this subject that is the focus of this book. To be sure, we include enough material on spherical and hyperbolic geometry to provide an informative counterpoint. In fact, we believe that a student who does not have a good background in Euclidean geometry is not in a position to get the point of non-Euclidean geometry.

It is well known that most students find it difficult to learn to do proofs. For most of the history of mathematics students studied geometry, especially Euclid, to learn about proofs. (Abraham Lincoln studied Euclid to improve his logic skills.) Our experience teaching convinces us that geometry provides an excellent setting for students to improve their proof skills. We've also found that the material adapts well to group learning.

There is a tremendous amount of Euclidean geometry beyond what is commonly taught in high schools. The first third of this book thoroughly covers secondary geometry course topics in a sophisticated manner. After covering this, we include material on the nine-point circle, the Fermat point, the Euler line, and other theorems from some of the world's greatest mathematicians. The results are not abstract, in the sense of modern mathematics, but many of them are deep and our approach is rigorous. We have attempted to create a text that satisfies the needs of prospective and current teachers, or, for that matter, any reader who appreciates powerful, elegant results. Our writing style is informal, and we have phrased proofs in the manner of present-day mathematicians.

It is reasonable to divide the book into three sections. Chapters 0 through 4 (with the exception of 3.4) cover material that very well may prove review for most students, although we doubt that most students will have seen real proofs of the material. The instructor may choose to spend a lot of time or just a little time on these sections. At any rate, chapters 0 through 5 (except for the enrichment sections 3.4 and 5.4) are basic for the rest of the book. Chapter 6, on regular polygons, is also for enrichment and may be safely omitted, if desired. Chapters 7 through 10 are all concerned with deeper properties of triangles. Although these are our favorite chapters, they are not necessary for the last chapters. Chapters 11 and 13 are also enrichment chapters. The main goal of the remaining three chapters chapters 12, 14 and 15, is to teach about non-Euclidean geometry in Chapter 15. In order to put it in some perspective, we first develop spherical geometry in Chapter 14. And, in order to develop the language of solid angles necessary for spherical geometry, we first present Chapter 12 on solid geometry. It is a surprising feature of elementary solid geometry that fairly simple statements—such as two lines parallel to the same line are parallel to each other—can be notoriously hard to prove. Again, the instructor can deal with this by spending a lot of time on it or by spending little time on it.

Some of the exercises are marked with stars to indicate that there is a hint in the back of the book. Of course, one of the great charms of proofs is that most theorems can be proven in many different ways; and we don't mean to say that the way we suggest in our hint is by any means the only right way to do this proof. We only mean to point the way to a proof which seems natural to us.

Thanks are due to several people who assisted us. We are grateful to the DePaul University Quality of Instruction Council for the support that led to a faster completion of the text than would otherwise have been possible. Our department administrative assistant, Nydia Rodriguez, put in many hours of typing, organizational help, and general assistance. This text could not have been produced without the cheerful and able help of Mark Jackowiak, who worked on all of the initial phases of the book, including figures, summaries and index, and who acted as a sounding-board for the authors. Georgia Katsis ably produced the original computer generated figures which were used by Network Graphics to produce the diagrams in the current text. We thank all of these individuals for their help. Finally, thanks are due to the many students with whom we have had the privilege and fun of sharing this information using our earlier notes, and who have helped us learn how to teach it.

Allan Berele, aberele@condor.depaul.edu
Jerry Goldman, jgoldman@condor.depaul.edu
DePaul University, Chicago

GEOMETRY
Theorems and Constructions

CHAPTER 0

Notation and Conventions

0.1 NOTATION

In this chapter we will introduce our basic terminology for geometric objects and the relations between them. Our main concern is the theorems of Euclidean geometry, and not the abstract construction of this geometry as an axiom system. We will *not* attempt to define terms such as points, lines, etc., and the relations between them, partly because we expect the reader to have already studied some geometry and to have a reasonable intuition about these objects, and partly because these objects don't have reasonable, honest definitions. There are some ambiguities in the conventions, but they are traditional and we will not attempt any modifications.

Points: Points will be denoted by capital letters.

Lines: The line containing the points A and B will be denoted $\overleftrightarrow{AB}$. We will also use script letters, especially $\ell's$, to denote lines. Three or more points that lie on the same line are said to be *collinear*. Three or more lines which meet at the same point are said to be *concurrent* at that point. Note that a line divides the plane into two *half planes*, each of which may be described by specifying the line and any one point in the half plane.

Rays: A ray is a "half line." The ray with initial point A and containing point B will be denoted $\overrightarrow{AB}$.

Line Segments: The line segment with end points A and B will be denoted $\overline{AB}$. If we leave the bar off and just write AB, we will mean the *length* of AB. So, for example, it makes sense to write $AB + CD$, but it does not make sense to write $\overline{AB} + \overline{CD}$, because we can add lengths but not line segments. A point P on $\overline{AB}$ that is distinct from A and B is called an interior point of $\overline{AB}$. P is the midpoint of $\overline{AB}$ if $PA = PB$.

Angles: There are two notations for angles. The symbol $\angle A$ refers to the angle with vertex at A. Although this notation is short, it may be ambiguous, because in a given figure there may be more than one angle with vertex at A. The symbol $\angle BAC$ refers to the angle with vertex A and sides $\overrightarrow{AB}$ and $\overrightarrow{AC}$. The same symbols $\angle A$ and $\angle BAC$ are also used to denote the sizes or measures of the angles! This is mildly unpleasant, but less pedantic and generally does not cause any problems in practice. We will use degrees rather than radians for angle measurements. A *right angle* is one whose measurement is $90°$; an *acute* angle has measure less than $90°$; while an *obtuse* angle has measure greater than $90°$. If $\angle A + \angle B = 90°$ then these angles are said to be *complements* of each other, whereas they are called *supplements* of each other if their

1

sum is 180°. The *interior* of $\angle ABC$ is the set of points common to the half plane on the C side of $\overleftrightarrow{AB}$ and the A side of $\overleftrightarrow{BC}$, excluding the points on $\overrightarrow{BA}$ and $\overrightarrow{BC}$. When P is interior to $\angle ABC$, then $\overrightarrow{BP}$ is a *bisector* of this angle if $\angle ABP = \angle PBC$.

Congruence: Two line segments of the same length are said to be congruent. We use the symbol $\cong$ to mean "is congruent to." So the statement $\overline{AB} \cong \overline{CD}$ means the same thing as the statement $AB = CD$. The latter is more traditional but we will generally write $\overline{AB} \cong \overline{CD}$ instead of $AB = CD$. This statement is *not* the same as the statement $\overline{AB} = \overline{CD}$, which means that $\overline{AB}$ and $\overline{CD}$ are the same line segment and so either $A = C$ and $B = D$ or $A = D$ and $B = C$.

We also say that two angles are congruent if they have the same size. Thanks to our ambiguous notation, the statement $\angle A \cong \angle B$ and the statement $\angle A = \angle B$ mean the same thing!

Triangles: The triangle with vertices A, B, and C is denoted $\triangle ABC$. An *equilateral triangle* has all three sides the same length. An *obtuse triangle* has an obtuse angle at some vertex, whereas an *acute triangle* has no obtuse or right angles. Finally, a *right triangle* has a 90° angle. The side opposite this angle is called the *hypotenuse*.

0.2 CONSTRUCTIONS

As in any advanced math book, our main concerns will be theorems and proofs. In addition, we will also be concerned with construction problems. In a construction problem, you are given certain geometric figures and asked how to construct other related figures from them. In this game it is presumed that you are allowed to use an unmarked straightedge and a compass. Or, equivalently, given any two points A and B you can draw $\overleftrightarrow{AB}$; moreover, given a point O and a segment $\overline{AB}$ you can draw a circle with center O and radius AB. The idea of using only circles and lines to do construction problems has its source in ancient Greek philosophy.

People have studied construction problems using different types of equipment. For example, there are theorems that state that a certain construction problem that cannot be solved using unmarked straightedge and compass can be solved if we allowed ourselves additional equipment, such as a straightedge we could mark. And, in the other direction, people have studied which constructions are possible if we use less equipment, such as a compass but no straightedge, or a two-sided straightedge, but no compass. In fact, classically it was assumed that compasses snapped shut when we picked them up so that we could construct a circle with center O and radius OA, but not with radius AB. Interesting articles on these subjects include "Euclidean construction and the geometry of Origami," by R. Geretschlagen in *Mathematics Magazine*, 1995(68), pp. 357–371; "A short elementary proof of the Mohr-Mascheroni theorem," by N. Hungerbühler in *The American Mathematical Monthly*, 1994(101), pp784–7; and "Geometric construction: The double straightedge," by W. Wenecker in *Mathematics Teacher*, 1971(69), pp. 697–704. It can be proven that any construction we can perform with our "modern" compasses can also be performed with these compasses which snap shut, thus there is almost no difference between the theory of constructions in this book and the classical theory in Euclid's *Elements of Geometry.*

We close with some free advice. We used to recommend that our students buy and use a ruler and a compass. This is not bad advise, but if you can buy or otherwise get access to the computer program Geometer's Sketchpad^{TM}, that is even better. Whichever you have, we recommend that you use them frequently as you read this book. The idea is that if you come across a construction of midpoints, or square roots, or whatever, you should stop and construct a few. Or, if you read a theorem that says the medians of a triangle meet in a point, you stop and draw a few triangles and their medians. Even for proofs, drawing your own diagrams can help you see what's going on.

CHAPTER 1

Congruent Triangles

1.1 THE THREE THEOREMS

The most basic relationship between two triangles in geometry is *congruence.* Here is the definition.

> **DEFINITION.** Two triangles triangle $\triangle$ABC and triangle $\triangle$DEF are said to be *congruent* if the following six conditions all hold:
>
> 1. $\angle A \cong \angle D$
> 2. $\angle B \cong \angle E$
> 3. $\angle C \cong \angle F$
> 4. $\overline{AB} \cong \overline{DE}$
> 5. $\overline{BC} \cong \overline{EF}$
> 6. $\overline{AC} \cong \overline{DF}$

In this case, we write $\triangle ABC \cong \triangle DEF$ and we say that the various congruent angles and segments "correspond" to each other. See Fig. 1.1.

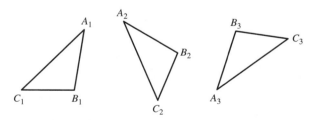

Figure 1.1: $\triangle A_1 B_1 C_1 \cong \triangle A_2 B_2 C_2 \cong \triangle A_3 B_3 C_3$

It is important to be careful with the notation. For example, if you write "$\triangle ABC$" or "$\triangle ACB$" you are describing the same triangle both times. However, the two names are not interchangeable when you are dealing with congruence. The statement "$\triangle ABC \cong \triangle DEF$" does not mean the same thing as the statement "$\triangle ACB \cong \triangle DEF$." For one thing, the first statement implies that $\overline{AB} \cong \overline{DE}$, and the second statement implies that $\overline{AB} \cong \overline{DF}$! (There are numerous other differences.)

You can see from the definition that if you are told that $\triangle ABC$ is congruent to $\triangle DEF$, you are being given a lot of information. Conversely, however, if you were given two triangles and you wanted to prove that they were congruent, then, from the definition, it seems that you would have a lot of work to do because you have six conditions to verify. Fortunately, there are three useful theorems about this question. Each says that if you know that certain of the conditions (1)–(6) were true, then these would imply that the others were also true and the triangles would be congruent. These theorems go by the names Angle-Side-Angle, Side-Angle-Side, and Side-Side-Side; or by the acronyms ASA, SAS, and SSS. Here are the statements of these theorems:

ASA Theorem. If $\angle A \cong \angle D$, $\angle B \cong \angle E$ and $\overline{AB} \cong \overline{DE}$, then $\triangle ABC \cong \triangle DEF$.

SAS Theorem. If $\angle A \cong \angle D$, $\overline{AB} \cong \overline{DE}$ and $\overline{AC} \cong \overline{DF}$, then $\triangle ABC \cong \triangle DEF$.

SSS Theorem. If $\overline{AB} \cong \overline{DE}$, $\overline{BC} \cong \overline{EF}$ and $\overline{AC} \cong \overline{DF}$ then $\triangle ABC \cong \triangle DEF$.

Before we turn to the proofs of these theorems, we point out two omissions from this list: AAA and SSA. This is because both are false! If $\triangle ABC$ and $\triangle DEF$ have their angles respectively congruent—$\angle A \cong \angle D$, $\angle B \cong \angle E$ and $\angle C \cong \angle F$—then the two triangles might or might not be congruent. In this case the two triangles are said

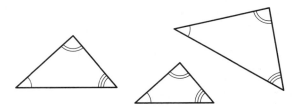

Figure 1.2: Similar triangles

to be *similar*, a relation we will study in Chapter 4. As you can see from Fig. 1.2, similar triangles need not be congruent. An example we will discuss later is afforded by equilateral triangles. If $\triangle ABC$ and $\triangle DEF$ are equilateral, then all six angles are 60° angles and so all are congruent. However, there are equilateral triangles of all different sizes that obviously are not congruent to each other. As for the falsity of SSA, see Fig. 1.3, and consider $\triangle ADC$ and $\triangle ADB$. Start off with triangle $\triangle ABC$ in which $\overline{AB} \cong \overline{AC}$ and extend $\overline{BC}$ to a point D. Each of $\triangle ADB$ and $\triangle ADC$ has $\angle D$ for one angle, and each has $\overline{AD}$ for one side. Since $\overline{AB} \cong \overline{AC}$, these triangles also have a second side congruent. So $\triangle ADB$ and $\triangle ADC$ satisfy the condition SSA but clearly are not congruent.

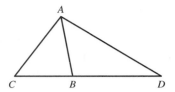

Figure 1.3: SSA is false

1.2   PROOFS OF THE THREE THEOREMS

Is it possible to prove SAS? When Euclid wrote his *Elements of Geometry* about 23 centuries ago, he included a proof of SAS as one of his very first theorems. When Hilbert wrote his *Foundations of Geometry* in 1899, he listed SAS among his axioms and did not prove it. This was not because Hilbert had not read Euclid's *Elements* and it was not because Hilbert couldn't understand Euclid's proof. And it wasn't because Hilbert found a mistake in Euclid's proof. It was because Hilbert (and modern mathematicians generally) had a different concept of what is or isn't a proof. In a sense, whether or not SAS is provable is a philosophical question. Here is a sketch of a "proof" of SAS of which a modern mathematician would not approve. Assume that

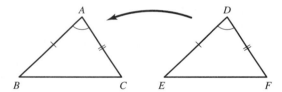

Figure 1.4: Proof of SAS

in $\triangle ABC$ and $\triangle DEF$, $\angle A \cong \angle D$, $\overline{AB} \cong \overline{DE}$, and $\overline{AC} \cong \overline{DF}$. Since $\angle A \cong \angle D$ we can lay $\angle D$ on top of $\angle A$ with side $\overrightarrow{DE}$ on $\overrightarrow{AB}$ and side $\overrightarrow{DF}$ on $\overrightarrow{AC}$. But since $\overline{DE} \cong \overline{AB}$, the point E will land on top of the point B. Likewise, since $\overline{DF} \cong \overline{AC}$, the point F will coincide with C. From this we can easily see that $\overline{EF} \cong \overline{AC}$ and likewise $\angle E \cong \angle B$ and $\angle C \cong \angle F$.

What is wrong with this proof? It is a convincing argument. An argument like this might well be used (perhaps with a scissors for dramatic effect) to explain SAS to a student. However, someone like Hilbert would complain that the proof is not "rigorous." We don't have an axiom system (certainly Euclid did not) that permits us to carry triangles from one part of the plane to another without deformation in order to lay them down on top of each other. The proof appeals to certain of our intuitive notions of congruence rather than precise unambiguous axioms. (If this makes you think that modern mathematics is a bit neurotic, you aren't the first one to think it. If you study chapter 15 on non-Euclidean geometry, you may learn a bit about how we got this way.)

We now turn to the proofs of ASA and SSS. These proofs will use SAS. The proof of SSS is long, so be patient with it.

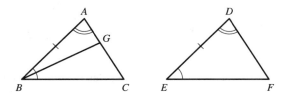

Figure 1.5: Proof of ASA

Proof of ASA. Assume that in $\triangle ABC$ and $\triangle DEF$ that $\angle A \cong \angle D$, $\angle B \cong \angle E$ and $\overline{AB} \cong \overline{DE}$. Consider now $\overline{AC}$ and $\overline{DF}$. If we knew that they were congruent, then we would be done: The two triangles would be congruent by SAS. So we assume that they are not equal and $\overline{AC}$ is longer. Since $\overline{AC}$ is longer there is an interior point G on $\overline{AC}$ such that $\overline{AG} \cong \overline{DF}$. But now $\triangle ABG$ and $\triangle DEF$ satisfy:

$$\angle A \cong \angle D,$$

$$\overline{AB} \cong \overline{DE}$$

and $\overline{AG} \cong \overline{DF}$.

So, by SAS, $\triangle ABG \cong \triangle DEF$. Hence,

$$\angle ABG \cong \angle E.$$

But, by hypothesis

$$\angle ABC \cong \angle E.$$

Comparing these two equations, we see that $\angle ABG \cong \angle ABC$. But this is impossible: $\angle ABG$ has to be smaller than $\angle ABC$ because G is an interior point of $\angle ABC$. This is a contradiction. Hence, our assumption that $\overline{AC}$ is not congruent to $\overline{DF}$ must be false. Therefore, $\triangle ABC \cong \triangle DEF$, and the proof is concluded. $\square$

In order to prove SSS we prove the following useful little theorem.

THEOREM. Let $\triangle ABC$ be any triangle.

1. If $\angle B \cong \angle C$ then $\overline{AB} \cong \overline{AC}$
2. If $\overline{AB} \cong \overline{AC}$ then $\angle B \cong \angle C$

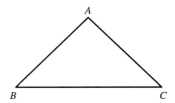

Figure 1.6: Isosceles triangle theorem

Proof. (1): In this case we have these three congruences,

$$\angle B \cong \angle C,$$

$$\angle C \cong \angle B,$$

$$\text{and } \overline{BC} \cong \overline{CB}.$$

Hence, $\triangle ABC \cong \triangle ACB$ by ASA (!). So $\overline{AB} \cong \overline{AC}$ as corresponding parts of congruent triangles. □

Proof. (2): Now use the congruences,

$$\overline{AB} \cong \overline{AC},$$

$$\overline{AC} \cong \overline{AB},$$

$$\text{and } \angle A \cong \angle A,$$

to conclude that $\triangle ABC \cong ACB$ by SAS. So $\angle B \cong \angle C$. □

A triangle with two congruent sides (angles) is called an *isosceles* triangle.

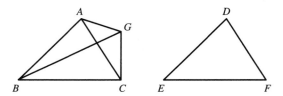

Figure 1.7: Proof of SSS

Proof of SSS. Assume in $\triangle ABC$ and $\triangle DEF$ that $\overline{AB} \cong \overline{DE}$, $\overline{BC} \cong \overline{EF}$ and $\overline{AC} \cong \overline{DF}$. We now construct a point G such that $\angle GBC \cong \angle E$ and $\angle GCB \cong \angle F$. By ASA, $\triangle GBC \cong \triangle DEF$. To prove the theorem we will prove that $\triangle GBC \cong \triangle ABC$. Construct $\overline{GA}$. Since $\overline{AB} \cong \overline{DE}$ by assumption, and $\overline{DE} \cong \overline{GB}$ as corresponding parts of congruent triangles, we conclude that $\overline{AB} \cong \overline{GB}$, by transitivity. Hence $\triangle BAG$ is an isosceles triangle.

So, we conclude from the previous theorem that

$$\angle BAG \cong \angle BGA.$$

By a similar line of reasoning $\triangle CAG$ is an isosceles triangle and

$$\angle CAG \cong \angle CGA.$$

Now consider the following:

$$\angle BAG > \angle CAG$$

$$\angle CAG \cong \angle CGA$$

$$\angle CGA > \angle BGA$$

$$\angle BGA \cong \angle BAG$$

Together, these imply that $\angle BAG > \angle$ BAG. This is impossible. Hence G must coincide with A, $\angle GBC = \angle B \cong \angle E$, and $\triangle ABC \cong \triangle DEF$ by SAS.   □

The preceeding proof contains errors. Look at it carefully. We made certain assumptions without knowing whether or not they were true. Can you find them? We assumed that the picture would look like Fig. 1.7 and that G would lie outside of $\triangle ABC$. We used that assumption to conclude that $\angle BAG > \angle CAG$ and $\angle CGA > \angle BGA$. What if G were inside the triangle? In order to complete the proof we must now consider that case. (See exercise 5 for a famous example of the kind of trouble you can get into with misleading figures.) If G is inside of $\triangle ABC$, as previously $\triangle BAG$ and

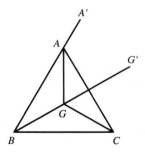

Figure 1.8: Proof of SSS (*cont.*)

$\triangle CAG$ are isosceles. Again, $\angle BAG \cong \angle BGA$ and $\angle CAG \cong \angle CGA$. Now extend $\overline{BA}$ to $\overline{BA'}$ and $\overline{BG}$ to $\overline{BG'}$, as shown in Fig. 1.8, and consider the exterior angle,

$$\angle A'AG = 180° - \angle BAG$$
$$= 180° - \angle BGA \text{ (since } \angle BAG \cong \angle BGA)$$
$$= \angle AGG'.$$

But

$$\angle A'AG > \angle CAG,$$
$$\angle CAG \cong \angle CGA,$$
$$\text{and } \angle CGA > \angle AGG'.$$

Hence, $\angle AGG' = \angle A'AG > \angle AGG'$ This contradiction shows that G cannot lie inside of $\triangle ABC$. As a final case we need to consider what would happen if G were on a side. We leave this last case as an exercise.

1.3 · APPLICATIONS TO CONSTRUCTIONS

We now apply our three triangle congruence theorems to solve a number of construction problems.

PROBLEM. To bisect an angle.

Let $\angle A$ be any angle. We want to construct a ray through A that will divide $\angle A$ into two congruent angles.

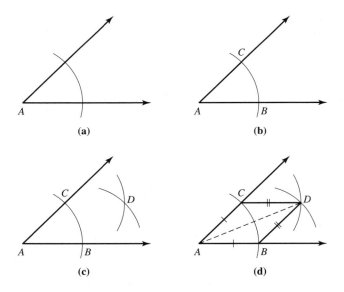

Figure 1.9: Construction of angle bisector

Solution. (See Fig. 1.9.) First, use your compass to construct B and C such that $\overline{AB} \cong \overline{AC}$. Then draw an arc with center B, and an arc with center C, both with the same radius. Let D be their point of intersection. Then $\overrightarrow{AD}$ is the angle bisector.

Proof. Since

$$\overline{AD} \cong \overline{AD},$$

$$\overline{AB} \cong \overline{AC},$$

and $\overline{BD} \cong \overline{CD}$,

we may conclude that $\angle ABD \cong \angle ACD$ by SSS. Hence, $\angle BAD \cong \angle CAD$.   □

PROBLEM. Given angle $\angle A$ and ray $\overrightarrow{BC}$, construct an angle $\angle DBC$ that will be congruent to $\angle A$.

Solution. (See Fig. 1.10.) Use your compass to draw a circle with center A and radius BC to find points P, Q on the two sides of $\angle A$ such that $\overline{AP} \cong \overline{AQ} \cong \overline{BC}$. Also, draw a circle with center B and radius BC. Now open the compass to length PQ and find a point D on this circle centered at B such that $\overline{PQ} \cong \overline{CD}$. (There will be two such points.) $\angle DBC$ will be the desired angle.

Proof. Since

$$\overline{AP} \cong \overline{BC},$$

$$\overline{AQ} \cong \overline{BD},$$

$$\overline{PQ} \cong \overline{CD},$$

we have $\triangle APQ \cong \triangle BCD$ by SSS. Hence $\angle A \cong \angle DBC$.   □

PROBLEM. Given a line ℓ and a point P not on ℓ, construct a line that passes through P and that is perpendicular to ℓ.

(Two lines that meet at a $90°$ angle are said to be perpendicular or orthogonal. We use the symbol $\perp$ for "is perpendicular to.")

Solution. (See Fig. 1.11.) Use your compass to find points A and B on ℓ such that $\overline{PA} \cong \overline{PB}$. Now, by drawing equal radius arcs centered at A and at B, find a point Q such that $\overline{AQ} \cong \overline{BQ}$. Then $\overleftrightarrow{PQ}$ is the desired line.

Proof. Consider the two triangles $\triangle PAQ$ and $\triangle PBQ$. They must be congruent by SSS. From this we conclude that

$$\angle APQ \cong \angle BPQ.$$

Now, let C be the point of intersection of $\overline{PQ}$ and $\overline{AB}$ and consider $\triangle PAC$ and $\triangle PBC$. These two triangles have $\overline{PC}$ as a common side, they have $\overline{PA} \cong \overline{PB}$ and they have congruent angles at P, as noted previously. Hence $\triangle PAC \cong \triangle PBC$. So, $\angle PCA \cong \angle PCB$. But, $\angle PCA + \angle PCB = 180°$. These two equations imply that $\angle PCA = 90°$ and $\angle PCB = 90°$, as claimed.   □

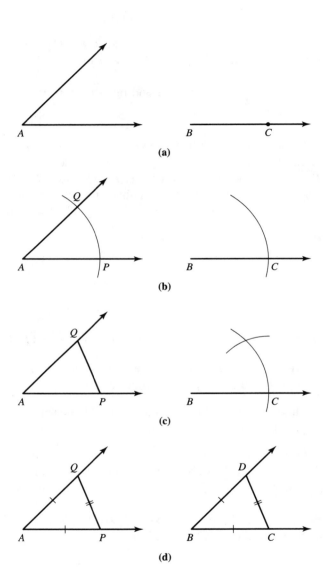

Figure 1.10: Copying an angle

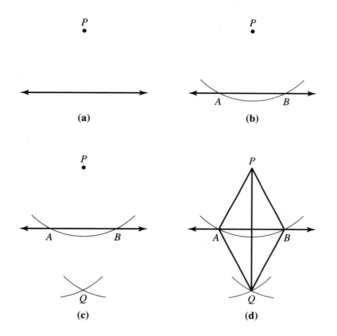

Figure 1.11: Construction of perpendicular from point to line

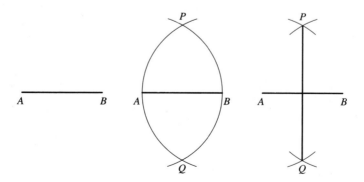

Figure 1.12: Construction of perpendicular bisector

PROBLEM. Given a line segment $\overline{AB}$, find a line that intersects $\overline{AB}$ at the midpoint and that is perpendicular to it. (This line is called the perpendicular bisector of $\overline{AB}$.)

Solution. With compass open to radius AB (or any convenient length greater than $\frac{1}{2}AB$), construct arcs with center at A and at B, meeting at P and Q. Then $\overleftrightarrow{PQ}$ will be the perpendicular bisector of $\overline{AB}$.

Proof. The proof is similar to the previous proof, and we leave it as an exercise for you.   □

1.4   APPLICATIONS TO INEQUALITIES

In Section 1.2 we proved that a triangle with two equal sides has two equal angles, and a triangle with two equal angles has two equal sides. Our main goal in this section is a generalization of this result; namely, we will describe what happens if we know that one side of a triangle is larger than another side, or that one angle is larger than another. Before proving the main result we first prove two theorems of independent interest, the vertical angle theorem and the exterior angle theorem.

THE VERTICAL ANGLE THEOREM. Let the lines $\overleftrightarrow{AB}$ and $\overleftrightarrow{CD}$ meet at point P, as shown in Figure 1.13 (P between A and B, and between C and D). Then $\angle APC \cong \angle BPD$ and $\angle APD \cong \angle BPC$.

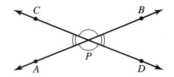

Figure 1.13: The vertical angle theorem

Proof. Clearly, $\angle APC + \angle APD = 180°$ and $\angle APD + \angle BPD = 180°$. Hence,

$$\angle APC = 180° - \angle APD$$
$$\angle BPD = 180° - \angle APD.$$

So $\angle APC \cong \angle BPD$. The case of $\angle APD \cong \angle BPC$ is similar.   □

THE EXTERIOR ANGLE THEOREM. In $\triangle ABC$ extend $\overline{BC}$ to a point D on $\overrightarrow{BC}$, forming the exterior angle $\angle ACD$. Then $\angle ACD$ is greater than each of $\angle A$ and $\angle B$, the remote interior angles.

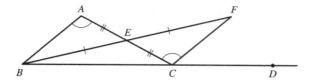

Figure 1.14: The exterior angle theorem

Proof. The first part of the proof will consist of constructing the picture in Fig. 1.14. First, let E be the midpoint of $\overline{AC}$; then connect B to E and extend to a point F in the interior of $\angle ACD$ such that $\overline{BE} \cong \overline{EF}$. Now consider the triangles $\triangle AEB$ and $\triangle CEF$. By construction, they have two of their respective sides congruent. Moreover, by the vertical angle theorem $\angle AEB \cong \angle CEF$. Hence $\triangle AEB \cong \triangle CEF$.

This implies that $\angle ECF \cong \angle A$. But $\angle ECF$ is less than $\angle ACD$. So $\angle A < \angle ACD$ as claimed. The case of $\angle B$ is similar.   □

The exterior angle theorem is equivalent to the statement that the sum of any two angles of a triangle is less than 180°. You probably know that in Euclidean geometry the sum of all three angles is 180°, and the exterior angle theorem is an easy consequence of that fact. However, since we have not yet proven the theorem about the sum of the three angles, we are not permitted to use it yet. Moreover, it is of some importance in the study of the foundations of geometry that the exterior angle theorem can be proved independently.

Here is the main result of this section.

THEOREM. Let $\triangle ABC$ be a triangle in which $\overline{BC}$ is longer than $\overline{AC}$. Then $\angle A > \angle B$. Or, the greater angle is opposite the greater side.

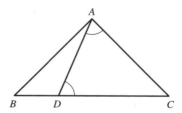

Figure 1.15: The Greater angle opposite the greater side

Proof. We may find a point D on $\overline{BC}$ such that $\overline{DC} \cong \overline{AC}$. Since $\triangle ADC$ is isosceles,

$$\angle CAD \cong \angle CDA.$$

By the exterior angle theorem

$$\angle ADC > \angle B,$$

and, clearly,

$$\angle A > \angle CAD.$$

Comparing these three statements, the result follows. □

The converse is now an easy consequence. (The converse of this particular theorem is about the same as the contrapositive.)

THEOREM. Let $\triangle ABC$ be a triangle in which $\angle A > \angle B$. Then $BC > AC$. Or, the greater side is opposite the greater angle.

Proof. The proof will be by contradiction. If the theorem were not true, then either $BC = AC$ or $AC > BC$. If $BC = AC$, then the triangle would be isosceles and we would get the contradiction $\angle A \cong \angle B$. If $AC > BC$, then by the previous theorem $\angle B > \angle A$, which is also a contradiction. □

We conclude this section and chapter with the famous "triangle inequality."

THEOREM. In any triangle the sum of the lengths of any two sides is greater than the length of the third side.

Proof. Let $\triangle ABC$ be any triangle. We will show that $AB + AC > BC$. Extend

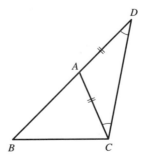

Figure 1.16: The sum of two sides greater than the third side

the side $\overline{BA}$ to a point D such that $\overline{AD} \cong \overline{AC}$. So $BD = AB + AC$. Now, since $\triangle ACD$ is isosceles, $\angle D \cong \angle DCA$, so $\angle D < \angle DCB$. So, in $\triangle BCD$, $\overline{BC}$ is a side opposite a smaller angle, $\angle D$, and $\overline{BD}$ is a side opposite a larger angle, $\angle DCB$. The theorem now follows. □

PROBLEMS

1.1. There are six conditions in the definition of congruent triangles, labeled (1), (2), (3), (4), (5), (6). Make a list of all (20) three-element subsets of $\{(1), ..., (6)\}$ and, for each one, tell whether it as hypothesis implies $\triangle ABC \cong \triangle DEF$.

*1.2. In our proof of SSS we choose G such that $\angle GBC \cong \angle E$, $\angle GCB \cong \angle F$ and such that G was on the same side of $\overleftrightarrow{BC}$ as A. Find a proof of SSS with G on the opposite side of $\overleftrightarrow{BC}$.

1.3. Prove SSA for right triangles: If $\overline{AB} \cong \overline{DE}$, $\overline{AC} \cong \overline{DF}$ and if $\angle B = 90°$ and $\angle E = 90°$ then $\triangle ABC \cong \triangle DEF$. (*hint:* Put $\angle B$ and $\angle E$ together to make a straight line.)

*1.4. Prove SAA: If $\angle A \cong \angle D$, $\angle B \cong \angle E$ and $\overline{BC} \cong \overline{EF}$, then $\triangle ABC \cong \triangle DEF$.

1.5. Find the mistake in the following "proof" that all triangles are isosceles, see Fig. 1.17. Let $\triangle ABC$ be any triangle. We will prove that $\overline{AB} \cong \overline{AC}$. Draw

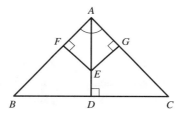

Figure 1.17: All triangles are isosceles

the perpendicular bisector of $\overline{BC}$ and the angle bisector of $\angle A$. Let these two lines intersect at the point E. From E draw lines $\overline{EF}$ and $\overline{EG}$ perpendicular to $\overline{AB}$ and $\overline{AC}$.

Now, $\triangle BED \cong \triangle CED$ by SAS, so $\overline{BE} \cong \overline{CE}$. Also, $\triangle AEF \cong \triangle AEG$ by SAA (see exercise 4), so $\overline{EF} \cong \overline{EG}$. Since $\overline{EF} \cong \overline{EG}$ and $\overline{BE} \cong \overline{CE}$, $\triangle BEF \cong \triangle CEG$ by SSA for right triangles (see exercise 3). From $\triangle AEF \cong \triangle AEG$ we conclude that $\overline{AF} \cong \overline{AG}$; from $\triangle BEF \cong \triangle CEG$ we conclude that $\overline{BF} \cong \overline{CG}$; and by addition we get $\overline{AB} \cong \overline{AC}$.

1.6. Let ℓ be a line and P a point on ℓ. Construct a line that contains P and that is perpendicular to ℓ.

1.7. Let ℓ be a line and P a point not on ℓ. Construct a line that contains P and that meets ℓ at a $45°$ angle. Construct a line that contains P and that meets ℓ at a $30°$ angle. (To do this exercise you need to use the fact that the sum of the angles in any triangle is $180°$.)

*1.8. Given triangles $\triangle ABC$ and $\triangle DEF$ such that $\overline{AB} \cong \overline{DE}$, $\overline{AC} \cong \overline{DF}$ and $\angle A > \angle D$, prove that $BC > EF$.

1.9. Given BC and $A_1, ..., A_n$, prove that $BC < BA_1 + A_1A_2 + A_2A_3 + ... + A_nC$.

1.10. This exercise extends the definition of congruence from triangles to more general polygons and explores some consequences. Let $2n$ points ($n \geq 3$, a whole number) $A_1, \ldots, A_n$ and $B_1, \ldots, B_n$ be given. Think of these points as vertices and $A_1 A_2 \ldots A_n$ as the polygon formed by the union of the segments $\overline{A_1 A_2}, \overline{A_2 A_3}, \ldots, \overline{A_n A_1}$. Define $A_1 A_2 \ldots A_n \cong B_1 B_2 \ldots B_n$ if for all distinct vertices A_i, A_j and A_k the two triangles $\triangle A_i A_j A_k$ and $\triangle B_i B_j B_k$ are congruent.

(a) Prove $ABCD \cong EFGH$ if $AB = EF$, $BC = FG$, $AD = EH$, $\angle A = \angle E$ and $\angle B = \angle F$.

(b) Prove $ABCD \cong EFGH$ if $AB = EF$, $BC = FG$, $CD = GH$, $AD = EH$ and $\angle A = \angle E$.

(c) Prove that two pentagons are congruent if SASASAS holds. [Hint: Use (a).]

(d) Prove that two polygons are congruent if all corresponding sides and angles are congruent. [Hint: Use induction on the number of sides.]

CHAPTER SUMMARY

- There are three basic triangle congruence theorems. These three and two additional sufficient conditions for congruence are:

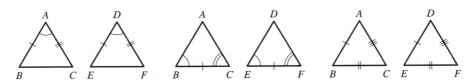

SAS: $\overline{AB} \cong \overline{DE}$, $\angle A \cong \angle D$, and $\overline{AC} \cong \overline{DF}$ implies that $\triangle ABC \cong \triangle DEF$.

ASA: $\angle B \cong \angle E$, $\overline{BC} \cong \overline{EF}$, and $\angle C \cong \angle F$ implies that $\triangle ABC \cong \triangle DEF$.

SSS: $\overline{AB} \cong \overline{DE}$, $\overline{BC} \cong \overline{EF}$, and $\overline{AC} \cong \overline{DF}$ implies that $\triangle ABC \cong \triangle DEF$.

- The two additional ones are:

AAS: $\angle B \cong \angle E$, $\angle C \cong \angle F$, and $\overline{AC} \cong \overline{DF}$ implies that $\triangle ABC \cong \triangle DEF$.

SSA: (For right triangles only!) $\overline{AB} \cong \overline{DE}$, $\overline{AC} \cong \overline{DF}$, and $\angle B \cong \angle E = 90°$ implies that $\triangle ABC \cong \triangle DEF$.

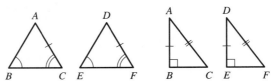

- Constructing an angle bisector:

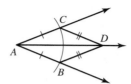

Find points C and B equal distances from A. Mark off a point D where two equidistant arcs centered at C and B intersect is the bisector.

- Copying an angle:

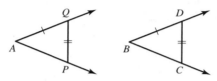

Open compass to convenient length $\overline{BC}$. Find points P and Q such that $\overline{AP} \cong \overline{BC} \cong \overline{AQ}$. Open compass to length $\overline{PQ}$ and strike-off point D on the other side of the angle such that $\overline{PQ} \cong \overline{CD}$. $\angle DBC$ is the desired angle.

- Constructing a perpendicular from a point to a line:

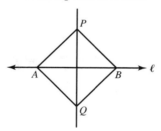

Find points A and B on ℓ such that $\overline{PA} \cong \overline{PB}$. Equal distance arcs centered at A and B intersect at point Q such that $\overline{AQ} \cong \overline{BQ}$. $\overleftrightarrow{PQ}$ is the desired line.

- Constructing a perpendicular bisector:

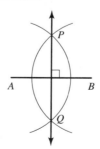

Open compass to convenient length (such as $\overline{AB}$) and construct arcs with centers at A and B, meeting at P and Q. $\overleftrightarrow{PQ}$ will be the perpendicular bisector of $\overline{AB}$.

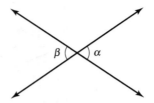

- Vertical angles β and α are congruent.

- If $\overline{BC}$ is longer than $\overline{AC}$, then $\angle A > \angle B$. (Greater angle is opposite the greater side) The converse is also true: If $\angle A > \angle B$, then $BC > AC$.

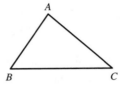

- In any triangle, the sum of the lengths of any two sides is greater than the length of the third side. For example,

$$AB + AC > BC.$$

- Exterior angle, $\angle ACD$ is greater than $\angle A$ and $\angle B$.

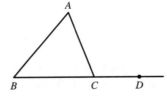

CHAPTER 2

Parallel Lines

2.1 EXISTENCE AND UNIQUENESS

One of the basic axioms of Euclidean geometry says that two points determine a unique line. This implies that two distinct lines cannot intersect in two or more points. They can either intersect in only one point or not at all. Two lines that don't intersect (i.e., that meet in zero points) are called parallel. Parallel lines will be the main subject of this chapter. We begin with their construction.

PROBLEM. Given a line ℓ and a point P not on ℓ, construct a line through P and parallel to ℓ.

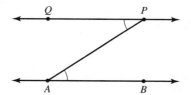

Figure 2.1: Construction of parallel lines

Solution. Let A be any point on ℓ, and draw $\overline{AP}$. Then draw a line $\overleftrightarrow{PQ}$ so that $\angle QPA \cong \angle PAB$ as shown in Fig. 2.1. This will be the desired line.

Proof. The proof will be by contradiction. If $\overleftrightarrow{PQ}$ and ℓ are not parallel we may assume without loss (as will be clear by the end of this proof) that they intersect as in Fig. 2.2, on the side of B at the point C.

Now consider $\triangle PAC$. The exterior angle $\angle APQ$ is equal to the interior angle $\angle PAC$. But this contradicts the exterior angle theorem, which states that $\angle QPA > \angle PAC$. Hence $\overleftrightarrow{PQ}$ must be parallel to ℓ. $\qquad\square$

We are now in a position to state two easily proven corollaries of this construction. (A corollary is a theorem that is a consequence of another theorem, construction, or proof.)

COROLLARY. Given a line ℓ and a point P not on ℓ, there exists a line that contains P and is parallel to ℓ.

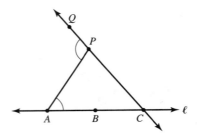

Figure 2.2: Proof that $\overleftrightarrow{PQ}$ and ℓ are parallel

This is what we have proven—that there exists at least one such line. We have not proven that it is the only one. For all we know at this point there could be two different lines, each of which contains P and each of which is parallel to ℓ.

COROLLARY. Given lines $\overleftrightarrow{AB}$ and $\overleftrightarrow{PQ}$ as in Fig. 2.1, if $\angle QPA \cong \angle PAB$, then $\overleftrightarrow{AB}$ is parallel to $\overleftrightarrow{PQ}$.

There is some notation and terminology that it is natural to introduce at this point. The symbol $\parallel$ is used to denote "is parallel to," and we write $\overleftrightarrow{PQ} \parallel \overleftrightarrow{AB}$ for $\overleftrightarrow{PQ}$ is parallel to $\overleftrightarrow{AB}$. Given any three lines ℓ_1, ℓ_2, ℓ_3, if ℓ_3 intersects both ℓ_1 and ℓ_2, then ℓ_3 is called a *transversal*. (This terminology is most commonly used when $\ell_1 \parallel \ell_2$, but it can be used in any case.) When $\overleftrightarrow{AB}$ and $\overleftrightarrow{PQ}$ are intersected by the transversal $\overleftrightarrow{AP}$ as in Fig. 2.1, then the pair of angles $\angle QPA$ and $\angle PAB$ are called opposite (or alternate) interior angles. So the preceding corollary can be restated as follows:

If two lines are cut by a transversal and if the opposite interior angles formed are congruent, then the two lines are parallel.

In order to continue our study of parallel lines, we need the famous parallel postulate. Our study of parallel lines up until now has dealt with their existence. The parallel postulate deals with the question of uniqueness, and we give it in a formulation due to John Playfair around 1795, which is equivalent to that given by Euclid.

THE PARALLEL POSTULATE. If ℓ is any line and P is a point not on ℓ, then there is no more than one line through P parallel to ℓ.

In light of our construction we can now conclude the following:

Given any line ℓ and a point P not on ℓ there exists exactly one line which contains P and is parallel to ℓ.

We can now prove the converse of our corollary.

THEOREM. Let $\overleftrightarrow{AB}$ and $\overleftrightarrow{PQ}$ be parallel lines with transversal $\overleftrightarrow{AP}$ such that $\angle QPA$ and $\angle PAB$ are opposite interior angles. Then $\angle QPA \cong \angle PAB$.

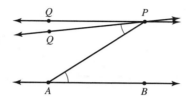

Figure 2.3: Proof that alternate interior angles are congruent

Proof. The proof will be by contradiction. If the theorem were false and if $\angle QPA \not\cong \angle PAB$, then we could construct a distinct line $\overleftrightarrow{PQ'}$ through P such that $\angle APQ' \cong \angle PAB$. Since $\angle APQ'$ and $\angle PAB$ are opposite interior angles, their congruence implies that $\overleftrightarrow{PQ'} \parallel \overleftrightarrow{AB}$. But this is now a contradiction of the parallel postulate: $\overleftrightarrow{PQ'}$ and $\overleftrightarrow{PQ}$ are two different lines, each goes through P and each is parallel to $\overleftrightarrow{AB}$. This contradiction comes about because we assumed that $\angle APQ \not\cong \angle PAB$. So these angles must be congruent.   □

2.2   APPLICATIONS

As the first application of our results on parallel lines we can prove that the sum of the angles of any triangle is 180°.

THEOREM. Let $\angle ABC$ be any triangle. Then $\angle A + \angle B + \angle C = 180°$.

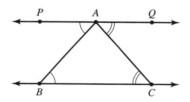

Figure 2.4: Sum of angles of triangles

Proof. Let $\overleftrightarrow{PAQ}$ be the line through A parallel to $\overleftrightarrow{BC}$ such that $\angle B$ and $\angle BAP$ are opposite interior angles and $\angle C$ and $\angle CAQ$ are opposite interior angles, as in Fig. 2.4. So $\angle B \cong \angle BAP$ and $\angle C \cong \angle CAQ$. Hence

$$\angle A + \angle B + \angle C = // \angle A + \angle BAP + \angle CAQ \qquad = 180°,$$

since $\angle A$, $\angle BAP$, and $\angle CAQ$ together make a straight line.   □

As a corollary we can calculate the sum of the angles of a quadrilateral.

COROLLARY. Let $ABCD$ be any quadrilateral. Then $\angle A + \angle B + \angle C + \angle D = 360°$.

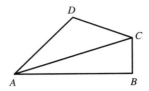

Figure 2.5: Sum of angles of quadrilateral

Proof. We draw the diagonal $\overline{AC}$, thus breaking the quadrilateral into two triangles. Note that $\angle A = \angle CAB + \angle CAD$ and $\angle C = \angle ACB + \angle ACD$. Hence,

$$\angle A + \angle B + \angle C + \angle D =$$

$$\angle CAB + \angle CAD + \angle B + \angle ACB + \angle ACD + \angle D =$$

$$(\angle ACD + \angle D + \angle DAC) + (\angle CAB + \angle B + \angle BCA).$$

The first sum of the last expression represents the sum of the angles of $\triangle ACD$ and the second sum represents the sum of the angles of $\triangle CAB$. Hence, each is $180°$ and together they add up to $360°$. $\qquad\square$

As a technical point we mention that if $ABCD$ is not convex it will have angles greater than $180°$. Some people don't define angles greater than $180°$. They would say that our corollary is only true for convex quadrilaterals.

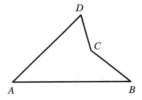

Figure 2.6: A non-convex quadrilateral

Another corollary is the SAA theorem. This theorem can be proved without using the fact that the sum of the angles of a triangle is $180°$ (cf: exercise 1.4). However, it is an easy consequence of that fact.

COROLLARY(SSA). In $\triangle ABC$ and $\triangle DEF$ assume that $\angle A \cong \angle D$, $\angle B \cong \angle E$, and $\overline{BC} \cong \overline{EF}$. Then $\triangle ABC \cong \triangle DEF$.

Proof. $\angle C = 180° - \angle A - \angle B$ and $\angle F = 180° - \angle D - \angle E$. Since $\angle A \cong \angle D$ and $\angle B \cong \angle E$, we conclude that $\angle C \cong \angle F$. Using the information $\angle B \cong \angle E$, $\angle C \cong \angle F$, and $\overline{BC} \cong \overline{EF}$, we see that $\triangle ABC \cong \triangle DEF$ by ASA. $\qquad\square$

Our next theorem gives another application of parallel lines to quadrilaterals.

THEOREM. Given a quadrilateral $ABCD$, the following are equivalent:

1. $\overleftrightarrow{AB} \parallel \overleftrightarrow{CD}$ and $\overleftrightarrow{AD} \parallel \overleftrightarrow{BC}$.
2. $\overline{AB} \cong \overline{CD}$ and $\overline{AD} \cong \overline{BC}$.
3. The diagonals bisect each other, i. e., if $\overline{AC}$ and $\overline{BD}$ intersect at E then $\overline{AE} \cong \overline{EC}$ and $\overline{BE} \cong \overline{ED}$.

In case you are not familiar with this type of statement, a " the following are equivalent"theorem is similar to an " if and only if"theorem. In our case it means that if you assume that hypothesis (1) is true, you can deduce (2) and (3); if you assume that hypothesis (2) is true, you can deduce (1) and (3); and if you assume (3) you can deduce (1) and (2). In our case we will prove that the theorem by proving statement (1) implies (2), (2) implies (1), (3) implies (2), and (2) implies (3). Although in principle this may not be the shortest way to prove our theorem, you can check that these four statements together do imply that (1), (2), and (3) are equivalent.

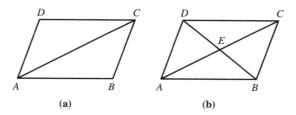

Figure 2.7: Parallelogram

Proof. Proof of (1) → (2): Draw the diagonal $\overline{AC}$, see Fig. 2.7(a). If we consider $\overline{AC}$ as transversal for the parallel lines $\overleftrightarrow{AB}$ and $\overleftrightarrow{CD}$, we see that $\angle BAC \cong \angle DCA$; and if we consider it as a transversal for the parallel lines $\overleftrightarrow{AD}$ and $\overleftrightarrow{BC}$ we see that $\angle BCA \cong \angle DAC$. Since $\overline{AC}$ is congruent to itself, $\triangle ABC$ and $\triangle CDA$ are congruent by ASA. Hence $\overline{AB} \cong \overline{CD}$ and $\overline{BC} \cong \overline{AD}$.

Proof of (2) → (1) : By assumption $\overline{AB} \cong \overline{CD}$ and $\overline{AD} \cong \overline{BC}$. Since $\overline{AC}$ is congruent to itself $\triangle ABC$ and $\angle CDA$ are congruent by SSS. Hence $\angle BAC \cong \angle DCA$. But these are alternate interior angles for the lines $\overleftrightarrow{AB}$ and $\overleftrightarrow{CD}$ cut by the transversal $\overline{AC}$. So $\overleftrightarrow{AB} \parallel \overleftrightarrow{CD}$. A similar argument shows that $\overleftrightarrow{AD} \parallel \overleftrightarrow{BC}$.

Proof of (2) → (3) : Since we have already proved that (2) → (1), we will assume that we know not only that opposite sides are congruent, but also that they are parallel. Hence, by alternate interior angles, $\angle DBA \cong \angle BDC$ and $\angle CAB \cong \angle ACD$ (see Fig. 2.7 (b)). Since $\overline{AB} \cong \overline{CD}$ we can conclude by ASA that $\triangle ABE \cong \triangle CDE$. So $\overline{AE} \cong \overline{EC}$ and $\overline{BE} \cong \overline{ED}$, as claimed.

Proof of (3) → (2) : We now assume that $\overline{AE} \cong \overline{EC}$ and $\overline{BE} \cong \overline{ED}$. Now $\angle AEB \cong \angle CED$ by the vertical angle theorem, so $\triangle AEB \cong \triangle CED$. Hence,

$\overline{AB} \cong \overline{CD}$. A similar argument shows that $\overline{BC} \cong \overline{AD}$.   □

A quadrilateral that satisfies these conditions is called a *parallelogram.* Two special cases of parallelograms are *rectangles* and *rhomboids.* A parallelogram in which all four angles are 90° is called a rectangle; a parallelogram in which all four sides are congruent is called a rhombus. For more properties of parallelograms, see the exercises.

2.3 DISTANCE BETWEEN PARALLEL LINES

LEMMA. Let ℓ be a line, P a point not on ℓ, and A and B distinct points on ℓ such that $\overline{PA}$ is perpendicular to ℓ. Then $PA < PB$.

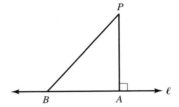

Figure 2.8: $\overline{PA}$ is the shortest segment from P to ℓ

Proof. In $\triangle PAB$, $\angle A = 90°$. How big is $\angle B$? Since the sum of the angles is 180°,

$$\angle B = 180° - \angle A - \angle P$$
$$= 90° - \angle P.$$

In particular, $\angle B < 90°$ and $\angle B < \angle A$. Since the longer side is opposite the larger angle, $PA < PB$, as claimed.   □

This lemma implies that of all the points on ℓ, A is the closest to P, and, of all the linear paths connecting P to ℓ, the line segment $\overline{PA}$ is the shortest. Given a line ℓ and a point P, we define the distance from P to ℓ to be the length of the perpendicular $\overline{PA}$. Our next theorem states that parallel lines are everywhere equidistant.

THEOREM. Let ℓ_1 and ℓ_2 be parallel lines and let P and Q be points on ℓ_2. Then the distance from P to ℓ_1 equals the distance from Q to ℓ_1.

Proof. Draw lines from P and from Q perpendicular to ℓ_1, meeting ℓ_1 at A and at B, respectively. Since $\angle PAB = 90°$ and $\angle QBA = 90°$, these angles are congruent. Moreover, $\angle QBA$ is congruent to the supplement of $\angle PAB$. So $\overline{PA} \| \overline{QB}$ by opposite interior angles. Similarly, $\overleftrightarrow{PQ} \| \overleftrightarrow{AB}$. Therefore, $PABQ$ must be a parallelogram, since opposite sides are parallel. Hence $\overline{PA} \cong \overline{QB}$, as claimed.   □

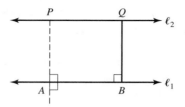

Figure 2.9: Parallel lines are everywhere equidistant

PROBLEMS

2.1. Let the lines ℓ_1 and ℓ_2 be cut by the transversal $\overleftrightarrow{AB}$ forming the eight angles a, b, c, d, e, f, g, h, as shown in Fig. 2.10. Prove that the following are equivalent:

(a) $\ell_1 \| \ell_2$

(b) $\angle a \cong \angle e$

(c) $\angle c \cong \angle g$

(d) $\angle b \cong 180° - \angle e$

(e) $\angle d \cong \angle h$

Note: The pairs a and e, b and f, c and g, d and h are called corresponding angles.

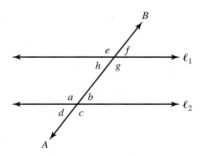

Figure 2.10: Exercise 1

2.2. Prove that the sum of the angles in a convex n-sided figure is $(n - 2)180°$.

*2.3. Given quadrilateral $ABCD$, prove that the following are equivalent:

(a) $ABCD$ is a parallelogram

(b) $\overline{AB} \| \overline{CD}$ and $\overline{AB} \cong \overline{CD}$

(c) $\angle A \cong \angle C$ and $\angle B \cong \angle D$

*2.4. Prove that a parallelogram $ABCD$ is a rectangle if and only if the diagonals $\overline{AC}$ and $\overline{BD}$ are congruent.

2.5. Prove that a parallelogram $ABCD$ is a rhombus if and only if the diagonals $\overline{AC}$ and $\overline{BD}$ are perpendicular to each other.

2.6. Given triangles $\triangle ABC$ and $\triangle DEF$ such that $\overline{AB}$ and $\overline{DE}$ are parallel and congruent and $\overline{BC}$ and $\overline{EF}$ are parallel and congruent, prove that $\overline{AC}$ and $\overline{DF}$ are parallel and congruent.

2.7. Given a quadrilateral $ABCD$ such that $\overline{AB} \parallel \overline{CD}$
   (a) Prove that $\angle C \cong \angle D$ if and only if $\overline{AD} \cong \overline{BC}$.
   (b) Prove that $\angle C > \angle D$ if and only if $AD > BC$.

2.8. Our proof that parallel lines are at a constant distance apart used the parallel postulate. (Can you see where?) Prove the following, without using the parallel postulate: Assume that lines ℓ_1 and ℓ_2 are at a constant distance apart. Then, if ℓ_1 and ℓ_2 are cut by a transversal, alternate interior angles must be congruent.

2.9. Euclid's postulate P.5, the famous "parallel postulate," states that if a straight line falling on twop straight lines makes the interior angles on the same side less than two right angles, the two straight lines, if produced indefinitely, meet on that side on which are the angles less than two right angles. Prove that Euclid's P.5 and Playfair's parallel postulate are equivalent (i. e. each implies the other).

CHAPTER SUMMARY

- The line through P such that $\angle APQ \cong \angle PAB$ will be parallel to ℓ. Moreover, if $\overleftrightarrow{QP}$ and $\overleftrightarrow{AB}$ are parallel and $\overleftrightarrow{PA}$ is a transversal, alternate interior angles, $\angle QPA$ and $\angle PAB$ are congruent.

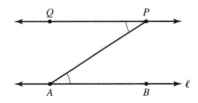

- The shortest distance between a point P and a line $\overleftrightarrow{AB}$ is measured along the line through P perpendicular to $\overleftrightarrow{AB}$.

- The sum of the measures of the angles of a triangle is $180°$.

  The sum of the measures of the angles of a quadrilateral is $360°$.

- Given quadrilateral ABCD, the following are equivalent:

  (1) Opposite sides are parallel.

(2) Opposite sides are congruent.

(3) The diagonals bisect each other.

When any one of the three conditions above holds, ABDC is a parallelogram.

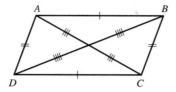

- Parallel lines are everywhere equidistant.

CHAPTER 3

Area

3.1 AREA OF RECTANGLES AND TRIANGLES

In this section we will discuss the formula giving the area of a rectangle as the product of its dimensions. Our study of area will be based on the following minimal assumptions:

For every polygon P we can associate to P a nonnegative real number denoted area (P) such that

(A1) Congruent polygons have equal areas.

(A2) If P can be cut into two non-overlapping polygons P_1 and P_2 then area$(P) =$ area$(P_1) +$ area(P_2).

(A3) If P is a square with sides of length 1, then area $(P) = 1$.

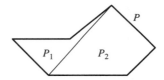

Figure 3.1: A polygon broken into two non-overlapping polygons

So far, the only congruent polygons we have discussed have been triangles, so it is not entirely clear how to apply (A1) in general. For our purposes, the only other case we will need is rectangles. Two rectangles are congruent if the bases are congruent and the heights are congruent. In general, two polygons $P_1 P_2 \cdots P_n$ and $Q_1 Q_2 \cdots Q_n$ are congruent if corresponding sides are congruent and corresponding angles are congruent, that is, $\overline{P_1 P_2} \cong \overline{Q_1 Q_2}, \ldots, \overline{P_n P_1} \cong \overline{Q_n Q_1}$ and $\angle P_1 \cong \angle Q_1, \ldots, \angle P_n \cong \angle Q_n$.

THEOREM. Let $ABCD$ and $EFGH$ be rectangles such that $\overline{AB} \cong \overline{EF}$ and $BC = \alpha \cdot FG$, for some positive real number α. Then area $(ABCD) = \alpha \cdot$ area $(EFGH)$.

Proof. To prove this theorem we must first prove the special case in which α is an integer, then the case in which α is a rational number, and finally the general case. We will only do the first two cases. Since the general case depends upon the properties of real numbers, we postpone it to the exercises in order not to cause difficulties for the student who has not studied the real number system.

If α is a positive integer, then the length of $\overline{FG}$ divides the length of $\overline{BC}$ evenly. Hence (cf. Fig. 3.2) we can find points $B_1, B_2, \ldots, B_{\alpha-1}$ between B and C on $\overline{BC}$ such that $\overline{FG} \cong \overline{BB_1} \cong \overline{B_1 B_2} \cong \cdots \cong \overline{B_{\alpha-1}C}$. At each point B_i draw a line segment parallel to $\overleftrightarrow{AB}$ connecting $\overline{BC}$ with $\overline{AD}$. This now breaks the rectangle $ABCD$ into α smaller non-overlapping rectangles. Each of these smaller rectangles has base congruent to $\overline{EF}$ and height congruent to $\overline{FG}$ and so, by (A1), each has area equal to area($EFGH$). But, by (A2), the area of $ABCD$ is the sum of the areas of these rectangles. So area($ABCD$) = $\alpha \cdot$ area($EFGH$). This proves the theorem for this special case.

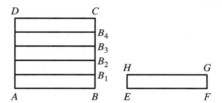

Figure 3.2: Case of α an integer ($\alpha = 5$)

We next consider the case in which α is a positive rational number. By definition, this means that $\alpha = \frac{n}{m}$ where n and m are integers. Now, given $ABCD$ and $EFGH$ we construct a third rectangle R. The base of R will be congruent to $\overline{AB}$ and to $\overline{EF}$. The height of R will be m times BC (Fig. 3.3). But since $BC = \alpha \cdot FG = \frac{n}{m} \cdot FG$, it follows that $n \cdot FG = m \cdot BC$. So the height of R is also n times FG. Since n and m are integers, we may apply the previous case. We have

$$\text{area}(R) = m \cdot \text{area}(ABCD)$$

and,

$$\text{area}(R) = n \cdot \text{area}(EFGH).$$

Hence,

$$m \cdot \text{area}(ABCD) = n \cdot \text{area}(EFGH).$$

Or, area $(ABCD) = \frac{n}{m}$ area $(EFGH) = \alpha\cdot$ area $(EFGH)$, which proves the theorem for rational α.

If α is not a rational number we can find rational numbers $\frac{n}{m}$ that approximate α very closely and apply the previous case. We will not supply the details here, but we refer you to exercise 14. $\qquad\square$

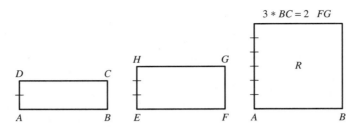

Figure 3.3: Case of α a rational number ($\alpha = \frac{2}{3}$)

COROLLARY. If $ABCD$ is a rectangle with $AB = 1$ and $CB = h$, then area $(ABCD) = h$.

Proof. Let $EFGH$ be a square, $EF = GH = 1$. By the theorem, area $(ABCD) = h \cdot$ area $(EFGH)$. But, by (A3), area $EFGH = 1$. □

COROLLARY. If $ABCD$ is a rectangle with $AB = b$ and $BC = h$, then area $(ABCD) = bh$.

Proof. Let $EFGH$ be a rectangle with $\overline{EF} \cong \overline{AB}$ and $FG = 1$ as in Fig. 3.4. By the theorem, area $(ABCD) = h \cdot$ area $(EFGH)$ and, by the previous corollary, area $(EFGH) = b$. □

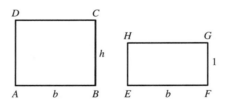

Figure 3.4: Area of a rectangle

We conclude this section by proving that the area of a triangle is one-half the length of its base times its height. Given $\triangle ABC$ we arbitrarily designate one side, say $\overline{BC}$, as the base. Then the height will be the distance from A to $\overleftrightarrow{BC}$. If D is a point on $\overleftrightarrow{BC}$ such that $\overleftrightarrow{AD} \perp \overleftrightarrow{BC}$, then the line segment $\overline{AD}$ is called the *altitude*. Of course, the length of $\overline{BC}$ will be the height.

THEOREM. The area of a triangle is $1/2 \times$ base $\times$ height.

Proof. Let $\triangle ABC$ be a triangle with base $\overline{BC}$. We will first prove the theorem in the special case in which $\angle B$ is a right angle. In this case $\overline{AB}$ is the altitude. Draw lines through A parallel to $\overleftrightarrow{BC}$ and through C parallel to $\overleftrightarrow{AB}$ meeting at

D (Fig. 3.5). The quadrilateral $ABCD$ has opposite sides parallel and it has a right angle at B. Hence it is a rectangle. So area $(ABCD) = AB \cdot BC$. But, area($ABCD$) is the sum of the areas of the two triangles, $\triangle ABC$ and $\triangle ADC$. Now, the diagonal of a rectangle (or any parallelogram) divides it into two congruent triangles, so $\triangle ABC \cong \triangle ADC$. In particular, area $ABC =$ area ADC. Combining all this information, we see

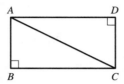

Figure 3.5: Area of a right triangle

$$AB \cdot BC = \text{area } ABCD$$

$$= \text{area } ABC + \text{area } ADC$$

$$= \text{area } ABC + \text{area } ABC$$

$$= 2 \cdot \text{area } ABC.$$

Thus, area $ABC = \frac{1}{2} \cdot AB \cdot BC$, as claimed for the case when $\angle B = 90°$.

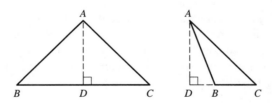

Figure 3.6: Area of Any Triangle

We now turn to the general case. Draw the altitude $\overline{AD}$ (Fig. 3.6) and compare the area of $\triangle ABC$ with the triangles $\triangle ABD$ and $\triangle ACD$. These latter two triangles are right triangles, and by the case we already considered

$$\text{area } ABD = \frac{1}{2}BD \cdot AD,$$

and

$$\text{area } ACD = \frac{1}{2}CD \cdot AD.$$

There are now two possibilities. If D is between B and C, then

$$\text{area } ABC = \text{area } ABD + \text{area } ACD$$
$$= \frac{1}{2}BD \cdot AD + \frac{1}{2}CD \cdot AD$$
$$= \frac{1}{2}(BD + CD) \cdot AD$$
$$= \frac{1}{2} \cdot BC \cdot AD.$$

If D is not between B and C, say it is past B, as shown, then

$$\text{area } ABC = \text{area } ACD - \text{area } ABD$$
$$= \frac{1}{2} \cdot CD \cdot AD - \frac{1}{2} \cdot BD \cdot AD$$
$$= \frac{1}{2}(CD - BD) \cdot AD$$
$$= \frac{1}{2} \cdot BC \cdot AD.$$

This completes the proof. $\qquad\qquad\qquad\qquad\qquad\qquad\qquad\qquad$ □

3.2   THE PYTHAGOREAN THEOREM

Let $\triangle ABC$ be a right triangle with right angle at C. Denote the lengths of the sides by $AB = c$, $AC = b$, and $BC = a$. Since $\angle C$ is the largest angle, $\overline{AB}$ will be the longest side. It is called the *hypotenuse*. The other two sides are called the legs. The Pythagorean theorem states that $a^2 + b^2 = c^2$, or, conversationally, the square of the hypotenuse equals the sum of the squares of the legs. As befits such a celebrated and venerable theorem, the Pythagorean theorem has many different proofs. We now present our favorite, which is based on areas.

THEOREM. Let $\triangle ABC$ be a right triangle with right angle at C. Denote $AB = c$, $BC = a$ and $AC = b$ as in Fig. 3.7. Then $a^2 + b^2 = c^2$.

Proof. As in Fig. 3.7, we construct a square $DEFG$ in which each side has length $a + b$. We next select points P_1, P_2, P_3, and P_4 on the sides such that $DP_1 = EP_2 = FP_3 = GP_4 = b$ and $P_1E = P_2F = P_3G = P_4D = a$. Then each of the four triangles $\triangle DP_1P_4$, $\triangle EP_2P_1$, $\triangle FP_3P_2$, and $\triangle GP_4P_3$ is congruent to $\triangle CAB$ by SAS. Hence, $P_1P_2 = P_2P_3 = P_3P_4 = P_4P_1 = c$, consequently, $P_1P_2P_3P_4$ is a rhombus. We claim that it is actually a square. To prove this we need to show that all of its angles are 90°. Consider, for example

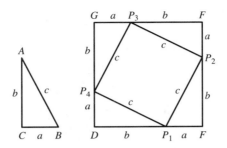

Figure 3.7: Proof of Pythagorean theorem

$\angle P_4 P_1 P_2$:

$$
\begin{aligned}
\angle P_4 P_1 P_2 &= 180° - \angle P_4 P_1 D - \angle P_2 P_1 E \\
&= 180° - \angle A - \angle B \\
&= \angle C \\
&= 90°.
\end{aligned}
$$

Therefore, $P_1 P_2 P_3 P_4$ is a square.

We now calculate the area of $DEFG$ in two different ways. On the one hand, since $DEFG$ is a square, it has area $(a + b)^2$. On the other hand, it has area equal to the sum of the areas of five pieces, the four triangles and the square. Each triangle has area $\frac{1}{2} \cdot ab$ and the square has area c^2. Hence

$$
\begin{aligned}
\text{area } EFGH &= (a + b)^2 \\
&= 4 \cdot \frac{1}{2} ab + c^2
\end{aligned}
$$

or

$$
(a + b)^2 = c^2 + 2ab.
$$

The theorem now follows easily.   □

Here are a few corollaries to this theorem:

COROLLARY (THE CONVERSE OF THE PYTHAGOREAN THEOREM). Let $\triangle ABC$ be any triangle and denote the lengths of the sides $AB = c$, $BC = a$, and $AC = b$. If $a^2 + b^2 = c^2$, then $\angle C = 90°$.

Proof. Construct a new triangle, $\triangle EFG$, such that $\angle G = 90°$, $\overline{EG} \cong \overline{AC}$ and $\overline{GF} \cong \overline{BC}$. Since $\triangle EFG$ is a right triangle we may apply the Pythagorean theorem to conclude that

$$
\begin{aligned}
EF^2 &= EG^2 + FG^2 \\
&= a^2 + b^2 \\
&= c^2.
\end{aligned}
$$

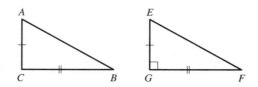

Figure 3.8: Converse of Pythagorean theorem

Hence, $EF = c$, which implies that $\overline{EF} \cong \overline{AB}$. Therefore, $\triangle ABC \cong \triangle EFG$ by SSS and $\angle G \cong \angle C = 90°$.   □

COROLLARY. In triangles $\triangle ABC$ and $\triangle EFG$ assume that $\overline{AB} \cong \overline{EF}$, $\overline{BC} \cong \overline{FG}$ and that $\angle C$ and $\angle G$ are right angles. Then $\triangle ABC \cong \triangle EFG$.

(This corollary is called SSA for right triangles, or the hypotenuse-leg theorem. It can be proved without resorting to the Pythagorean theorem, as we pointed out in Exercise 3 in Chapter 1; but this proof is fairly easy and gives a nice application of the Pythagorean theorem.)

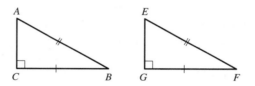

Figure 3.9: SSA theorem for right triangles

Proof. By the Pythagorean theorem

$$AC^2 + BC^2 = AB^2,$$

so

$$AC^2 = AB^2 - BC^2.$$

Likewise, $EG^2 = EF^2 - FG^2$. Since $AB = EF$ and $BC = FG$, it follows that $AC^2 = EG^2$ or $AC = EG$. Hence, $\triangle ABC \cong \triangle EFG$ by SSS.   □

Our third application of the Pythagorean theorem is called the geometric law of cosines. We have shown that $c^2 = a^2 + b^2$ if and only if $\angle C = 90°$. The geometric law of cosines relates c^2 to $a^2 + b^2$ even if $\angle C$ is not a right angle.

THEOREM. Let $\triangle ABC$ be a triangle with base $\overline{BC}$ and altitude $\overline{AD}$. Denote $AB = c$, $BC = a$, $AC = b$, and $CD = x$. If D is between B and C or if B is between C and D, then $c^2 = a^2 + b^2 - 2ax$. If C is between B and D, then $c^2 = a^2 + b^2 + 2ax$.

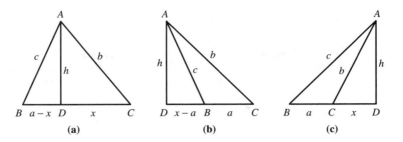

Figure 3.10: Geometric law of cosines

Proof. Consider the case (a) of Fig. 3.10, in which D is on $\overline{BC}$. Write h for AD. Since $\triangle ADC$ and $\triangle ADB$ are right triangles we have

$$h^2 + x^2 = b^2 \text{ and } h^2 + (a-x)^2 = c^2.$$

Hence

$$\begin{aligned} b^2 - x^2 &= c^2 - (a-x)^2 \\ &= c^2 - a^2 + 2ax - x^2. \end{aligned}$$

The result now follows using a bit of algebra. The other two cases are similar, and we leave their proofs to you.   □

3.3 AREA OF TRIANGLES

In this section we prove two deeper results on areas of triangles. The first compares the areas of two different triangles sharing one equal angle and the second gives the area of a triangle in terms of the lengths of the three sides.

THEOREM. Let $\triangle ABC$ and $\triangle DEF$ be triangles in which $\angle A \cong \angle D$. Then

$$\frac{\text{area } \triangle ABC}{\text{area } \triangle DEF} = \frac{AB \cdot AC}{DE \cdot DF}.$$

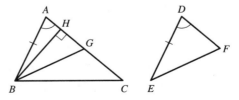

Figure 3.11: Triangles with an equal angle and an equal including side

Proof. We first consider the special case in which $\overline{AB} \cong \overline{DE}$. If $\overline{AC} \cong \overline{DF}$, then $\triangle ABC \cong \triangle DEF$ and the relation of the theorem reduces to $1 = 1$. We lose no generality by supposing that $AC > DF$. Choose a point G on $\overline{AC}$ such that $\overline{AG} \cong \overline{DF}$. Now, $\triangle ABG \cong \triangle DEF$ by SAS. Let $\overline{BH}$ be the altitude from B to $\overline{AC}$. Then area$(ABC) = \frac{1}{2} \cdot BH \cdot AC$ and area$(ABG) = \frac{1}{2} BH \cdot AG$. Hence

$$\frac{\text{area}(ABC)}{\text{area}(DEF)} = \frac{\text{area }(ABC)}{\text{area}(ABG)}$$

$$= \frac{\frac{1}{2} BH \cdot AC}{\frac{1}{2} \cdot BH \cdot AG}$$

$$= \frac{AC}{AG}$$

$$= \frac{AC}{DF}.$$

Since we assume that $AB = DE$, this last ratio is also equal to $\frac{AC \cdot AB}{DF \cdot DE}$. This completes the proof in this special case.

The general case now follows easily from a clever construction. Given two

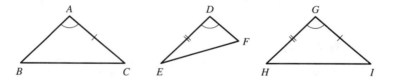

Figure 3.12: General case of triangles with equal angle

triangles $\triangle ABC$ and $\triangle DEF$ such that $\angle A \cong \angle D$, we construct a third triangle $\triangle GHI$ such that $\angle G$ is congruent to $\angle A$ and $\angle D$, $\overline{GH} \cong \overline{DE}$, and $\overline{GI} \cong \overline{AC}$. We may now apply the special case twice to the triangles $\triangle ABC$ and $\triangle GHI$ and to $\triangle DEF$ and $\triangle GHI$ to yield the relations

$$\frac{\text{area }(ABC)}{\text{area }(GHI)} = \frac{AB \cdot AC}{GH \cdot GI}$$

and

$$\frac{\text{area }(GHI)}{\text{area }(DEF)} = \frac{GH \cdot GI}{DE \cdot DF}.$$

Multiplying the first equation by the second yields the result. $\qquad\square$

Our last theorem in this chapter is called Heron's formula. It computes the area of a triangle in terms of the lengths of the sides.

HERON'S THEOREM. Let $\triangle ABC$ be any triangle and denote the lengths of the sides by $AB = c$, $BC = a$, and $AC = b$. Also, let $s = $ the semiperimeter, $s = \frac{1}{2}(a + b + c)$. Then the area of $\triangle ABC =$

$$\sqrt{s(s - a)(s - b)(s - c)}.$$

Proof. Let $\overline{BC}$ be the base and $\overline{AD}$ the altitude. As in the geometric law of cosines, let $AD = h$ and $CD = x$ (Fig. 3.13). Then, by that law,

$$c^2 = a^2 + b^2 - 2ax$$

so

$$x = \frac{a^2 + b^2 - c^2}{2a}.$$

(For the sake of simplicity we will assume that D is between B and C.) Also, by the Pythagorean theorem,

$$h^2 + x^2 = b^2$$

so

$$h^2 = b^2 - x^2.$$

Combining these two equations yields

$$h^2 = b^2 - \left(\frac{a^2 + b^2 - c^2}{2a}\right)^2.$$

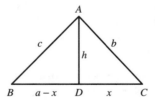

Figure 3.13: Heron's Formula

In principle, this is already enough information to find the area: Simply take a square root and multiply by $\frac{1}{2}a$. The main tool for the simplification will be the formula for the difference of two squares, namely, $x^2 - y^2 = (x - y)(x + y)$.

After one application of that rule we get

$$h^2 = b^2 - \left(\frac{a^2 + b^2 - c^2}{2a}\right)^2$$

$$= \left(b - \frac{a^2 + b^2 - c^2}{2a}\right)\left(b + \frac{a^2 + b^2 - c^2}{2a}\right)$$

$$= \left(\frac{2ab - a^2 - b^2 + c^2}{2a} \right) \left(\frac{2ab + a^2 + b^2 - c^2}{2a} \right).$$

Now multiply both sides by a^2 to see that

$$a^2 h^2 = \frac{1}{4}(2ab - a^2 - b^2 + c^2)(2ab + a^2 + b^2 - c^2) = \frac{1}{4}[c^2 - (a-b)^2][(a+b)^2 - c^2].$$

Again, each factor is the difference of two squares, so

$$a^2 h^2 = \frac{1}{4}(c - a + b)(c + a - b)(a + b + c)(a + b - c).$$

Finally, we divide by 4:

$$\frac{1}{4} a^2 h^2 = \frac{c - a + b}{2} \cdot \frac{c + a - b}{2} \cdot \frac{a + b + c}{2} \cdot \frac{a + b - c}{2}.$$

But $\frac{c-a+b}{2} = s - a$, $\frac{c+a-b}{2} = s - b$, $\frac{a+b+c}{2} = s$, and $\frac{a+b-c}{2} = s - c$, so the right hand side of the equation equals $s(s - a)(s - b)(s - c)$. The left-hand side is $(\frac{1}{2}ah)^2$, which is the square of the area. If we take the square root of both sides we get Heron's formula. $\qquad \square$

Our proof of Heron's formula uses a lot of algebra. Amazingly, Heron proved his formula long before symbolic algebra was invented. You may like to read about Heron's "nonalgebric" proof in B. M. Olive's article "Heron's remarkable triangle area formula," in *Mathematics Teacher*, Feb. 1993.

3.4   CUTTING AND PASTING

Informally, P and Q are *equidecomposible* if you can cut P into pieces and rearrange the pieces to form Q. Here is the more precise definition.

> **DEFINITION.** Two polygons P and Q are said to be equidecomposible if P can be written as a union of non-overlapping polygons $P_1, \ldots, P_n$ and Q can be written as a union of non-overlapping polygons $Q_1, \ldots, Q_n$ such that $P_1 \cong Q_1, \ldots, P_n \cong Q_n$ (Fig. 3.14). We will also say that P is equidecomposible with Q, or more simply that P is decomposed to Q.

It follows from the area axioms (A1) and (A2) that if two polygons are equidecomposible, then they have the same area. So equidecomposibility can be thought of as a more pristine approach to area that does not make use of numbers. Here is the surprising theorem:

> **THEOREM.** If two polygons P and Q have equal area, then they are equidecomposible.

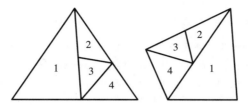

Figure 3.14: Equidecomposible triangle and quadrilateral

The following lemma is the key result in proving the theorem. We will explain and illustrate it but not give a completely formal proof. It says that equidecomposibility is a transitive relation.

LEMMA. Given polygons P, Q, and R such that P and Q are equidecomposible and Q and R are equidecomposible, then P and R are equidecomposible.

If we think of equidecomposibility in terms of cutting up a polygon and rearranging the pieces, then the lemma is clear: In order to decompose P into R, we first cut P up and rearrange the pieces to form Q, and then cut Q up and rearrange the pieces to form R. In terms of the formal definition using congruent polygons, the situation is about the same. There are two decompositions of Q, one which can be rearranged to form P and one of which can be rearranged to form R. If we superimpose these two decompositions of Q, we get one that can be rearranged to form either P or R (Fig. 3.15).

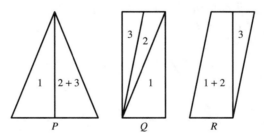

Figure 3.15: Equidecomposibility is transitive

For the rest of the proof of the theorem, pick out any line segment $\overline{XY}$. Say for convenience that $\overline{XY}$ has length 1. We will prove that any polygon P is equidecomposible with a rectangle with base congruent to $\overline{XY}$. This will imply that if P and Q have equal area, then they are each equidecomposible with the same rectangle and, in light of the lemma, this will imply that they are equidecomposible with each other.

We first consider the case of triangles. We want to show that any triangle is equidecomposible with a rectangle of base 1. We do this using a series of constructions. The first three show how to do it in the special case that things work out nicely, and then the next two show what adjustments can be made if they don't.

CONSTRUCTION 1. Decompose any triangle with a parallelogram.

Solution. Let $\triangle ABC$ be any triangle. Let D be the midpoint of $\overline{AC}$, $\overleftrightarrow{DE}$ parallel to $\overleftrightarrow{BC}$ and $\overleftrightarrow{CF}$ parallel to $\overleftrightarrow{AB}$, as shown in Fig. 3.16. Then $BEFC$ is a parallelogram, since opposite sides are parallel, and $\triangle DCF \cong \triangle DAE$ by ASA. So $\triangle ABC$ is equidecomposible with $BEFC$.

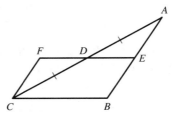

Figure 3.16: Any triangle decomposed to a parallelogram

CONSTRUCTION 2. Decompose a parallelogram with a parallelogram with one side of length 1, if things work out nicely.

Solution. Let $ABCD$ be the parallelogram. If we're lucky, then there is a point E on $\overline{DC}$ with $AE = 1$. In this case, let F on $\overrightarrow{EC}$ be such that $\overline{EF} \cong \overline{AB}$ (Fig. 3.17). Then $AEFB$ must be a parallelogram because the sides $\overline{EF}$ and $\overline{AB}$ are parallel and congruent. Finally, $\triangle ADE \cong \triangle BCF$ by ASA: $\angle D \cong \angle BCF$ because $\overline{AD} \parallel \overline{BC}$, $\angle AED \cong \angle F$ because $\overline{AE} \parallel \overline{BF}$, and $DE = DC - EC = EF - EC = CF$.

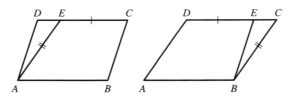

Figure 3.17: A parallelogram decomposed to a parallelogram with one side $= 1$

In the preceeding construction, there are two ways that things might not work out nicely. One such way would be if the height of $ABCD$ is less than 1; then there will be no point E on $\overrightarrow{DC}$ with $AE = 1$. In this case we can use construction 4 (which follows) one or more times to decompose $ABCD$ with a parallelogram with height less than or equal to 1. The other possible problem is if the points E on $\overleftrightarrow{DC}$ with $AE = 1$ are all to the left or right of $\overline{DC}$. In this case, we can use construction 5 (which follows) to decompose $ABCD$ with a parallelogram with the same base $\overline{AB}$ and the same height, but with opposite side further to the right or left as required.

So the missing step in this part of the program is to make our parallelogram into a rectangle.

CONSTRUCTION 3. Decompose a parallelogram with one side equal to 1 to a rectangle with one side equal to 1, if we're lucky.

Solution. Let $ABCD$ be the parallelogram with $AB = 1$. Say that $\angle A$ is obtuse, and let the altitude from A be $\overline{AE}$ and the altitude from B be $\overline{BF}$. If things work out nicely, then E will be on $\overline{CD}$. In this case we are done because $\triangle AEC \cong \triangle BFD$. We leave the proof of this congruence to you. If things don't work out

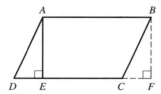

Figure 3.18: A parallelogram decomposed to rectangle

nicely, E will be to the right or left of $\overline{CD}$. In this case, we can move $\overline{CD}$ using construction 5.

Here now are the two constructions we need for the unlucky cases.

CONSTRUCTION 4. Decompose a parallelogram with a parallelogram with half the height.

Solution. See Fig. 3.19.

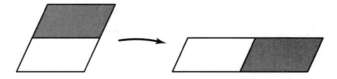

Figure 3.19: A parallelogram decomposed to a shorter parallelogram

CONSTRUCTION 5. Decompose any parallelogram $ABCD$ with a parallelogram $ABC'D'$ with $C' = D$ or $C = D'$.

Solution. We've already proven that the diagonal of a parallelogram divides it into two congruent triangles. With Fig. 3.20 for a hint you should be able to see how the decomposition works.

Proof of Main Theorem. Given any polygon P, we may cut P into non-overlapping triangles $T_1, \ldots, T_n$. Using constructions 1–5, we can decompose each of $T_1, \ldots$

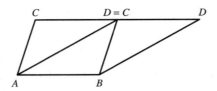

Figure 3.20: Shifting a parallelogram

., T_n into a rectangle $R_1,$. . .,R_n, such that each of these rectangles has base 1. Then these rectangles can be stacked to form a bigger rectangle R, also of base 1. So, any polygon can be decomposed to a rectangle of base 1. As we remarked earlier, the transitive property of equidecomposibility now implies that any two polygons with equal area are equidecomposible. ☐

PROBLEMS

3.1. Let $ABCD$ be a parallelogram. Define the base to be $\overline{AB}$ and the height to be the distance between $\overleftrightarrow{AB}$ and $\overleftrightarrow{CD}$. Prove that ABCD has area = base × height.

3.2. Assume that in quadrilateral $ABCD$ that $\overleftrightarrow{AB}\|\overleftrightarrow{CD}$. (Such a figure is called a *trapezoid*.) Let $AB = b_1$, $CD = b_2$ and let $h = the$ distance between $\overleftrightarrow{AB}$ and $\overleftrightarrow{CD}$.. Prove that $ABCD$ has area $\frac{1}{2}h(b_1 + b_2)$.

3.3. In Fig. 3.21, $ABCD$ is a rectangle and E lies on $\overline{AC}$. Prove that rectangle I and rectangle II have the same area. Use this fact to solve the construction problem: Given a rectangle R and a line segment $\overline{AB}$, construct a rectangle $ABCD$ with area equal to the area of R.

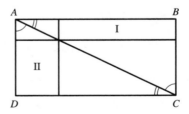

Figure 3.21: Exercise 3

3.4. Use exercise 3 to solve this construction problem: Given two rectangles R_1 and R_2 find a rectangle R such that area $R = $ area $R_1 + $ area R_2. Can you do it for more than two rectangles?

*3.5. Prove that in a right triangle, if the hypotenuse is the base of length c, then the height is $h = \frac{ab}{c}$.

*3.6. Given a convex quadrilateral $ABCD$ with $AC \perp BD$, prove that $AB^2 + CD^2 = BC^2 + AD^2$.

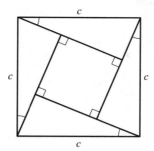

Figure 3.22: Exercise 7. Proof of Pythagorean theorem

3.7. Give another proof of the Pythagorean theorem based on Fig. 3.22.

3.8. Fig. 3.23 shows a sketch of Euclid's proof of the Pythagorean theorem. Fill in the details.

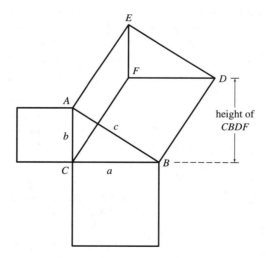

Figure 3.23: Exercise 8. Proof of Pythagorean Theorem

Construct F such that $\overleftrightarrow{EF} \parallel \overleftrightarrow{AC}$ and $\overleftrightarrow{DF} \parallel \overleftrightarrow{BC}$. Then $c^2 =$ area of $ACBDFE =$ area of $CBDF +$ area $AEFC$. Note that the area of $CBDF = CB^2$, since it is a parallelogram with base $\overline{BC}$ and height BC. Likewise, $AEFC$ has area AC^2.

3.9. Let $\triangle ABC$ be such that $\angle C = 90°$, $\angle A = 60°$, and $\angle B = 30°$. Prove that $AB = 2AC$ and $BC = \sqrt{3}AC$.

3.10. **(a)** Given $\overline{AB}$ and $\overline{CD}$ construct $\overline{EF}$ such that $EF^2 = AB^2 + CD^2$.

　　　　(b) Given $\overline{AB}$ and $\overline{CD}$ such that $AB \geq CD$, construct $\overline{EF}$ such that $EF^2 = AB^2 - CD^2$.

    (c)   Given $\overline{AB}$, $\overline{CD}$ and $\overline{EF}$ construct $\overline{GH}$ such that $GH^2 = AB^2 + CD^2 + EF^2$.

3.11. Prove the analogue of the Pythagorean theorem that uses equilateral triangles instead of squares.

*3.12. Let $\triangle ABC$ and $\triangle DEF$ be such that $\angle A$ and $\angle D$ are supplementary (i.e., $\angle A + \angle D = 180°$). Prove that

$$\frac{\text{area }(ABC)}{\text{area }(DEF)} = \frac{AB \cdot AC}{DE \cdot DF}$$

3.13. **(a)**   Assume that in $\triangle ABC$, $a = 4$, $b = 9$, and $c = 11$. Calculate the area.

    (b)   Calculate the length of each of the three altitudes in the triangle of part (a).

    (c)   If we let $a = 2$, $b = 3$, and $c = 7$ in Heron's formula, we get a problem. What is this problem and why does it happen?

3.14. In our proof of the first theorem in this chapter we omitted the last case of α any positive real number. Complete the proof using the following property of real numbers: Let $x < y$ be any two real numbers. Then there exists a rational number $\frac{n}{m}$ such that $x < \frac{n}{m} < y$.

[*Hint*: Use a proof by contradiction. First assume that area $(ABCD) < \alpha \cdot$ area $(EFGH)$. By the property of real numbers we can find $\frac{n}{m}$ such that

$$\alpha > \frac{n}{m} > \frac{\text{area } ABCD}{\text{area } (EFGH)}.$$

Construct a rectangle R with base $\overline{AB}$ and height $\frac{n}{m} \cdot FG$. Compare the area of R with that of $ABCD$ and $EFGH$. What is the contradiction? Next, do the case of area $(ABCD) > \alpha \cdot$ area $EFGH$.] *Remark*: This property of the real numbers can be expressed by saying that the rational numbers are dense in the reals. The application of this fact to geometry was known by the ancient Greeks. It is the basis of the method of exhaustion, and it plays a role in geometry similar to the role of limits in calculus.

3.15. Show how each of these pairs of polygons can be decomposed into congruent polygons:

    (a)   A 1×1 square and a $\frac{2}{3} \times \frac{3}{2}$ rectangle

    (b)   An isosceles right triangle and a square

    (c)   A 1×1 square and a rectangle with one side equal to $\sqrt{2}$

    (d)   An equilateral triangle with each side of length 1 and an isosceles triangle with base of length $\frac{1}{2}$

CHAPTER SUMMARY

- Area assumptions:

    (A1) Congruent polygons have equal areas.

    (A2) If P_1 and P_2 are two non-overlapping polygons, then the area of the polygon P they form together = area (P_1) + area (P_2).

    (A3) If P is a square with sides of length 1, then the area of $P = 1$.

- If $ABCD$ is a rectangle with $AB = b$ and $BC = h$, then area $(ABCD) = bh$.

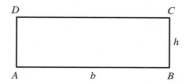

- The area of a triangle is $\frac{1}{2}bh$.

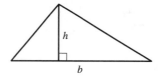

- If $\triangle ABC$ is a right triangle, the Pythagorean Theorem states that $AB^2 + BC^2 = AC^2$.

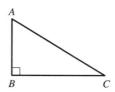

    The converse of the Pythagorean theorem is also true. Given the above equation in $\triangle ABC$, then $\angle B$ must be equal to $90°$.

- These two relationships comprise the geometric law of cosines:

$$c^2 = a^2 + b^2 - 2ax \text{ in cases (a) and (b), and}$$

$$c^2 = a^2 + b^2 + 2ax \text{ in case (c).}$$

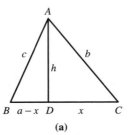 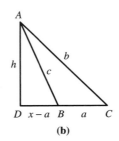 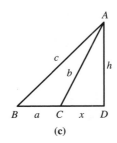

- Given $\triangle ABC$ and $\triangle DEF$ such that $\angle A \cong \angle D$, then

$$\frac{\text{area }(\triangle ABC)}{\text{area }(\triangle DEF)} = \frac{AB \cdot AC}{DE \cdot DF}$$

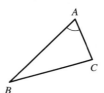

- The area of a triangle by Heron's formula is

$\sqrt{s(s-a)(s-b)(s-c)}$ where s is the semiperimeter $= \frac{1}{2}(a+b+c)$.

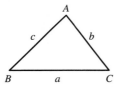

- If two polygons P and Q have the same area, it is possible to decompose P into smaller pieces and rearrange these pieces to form Q. This means that polygons with equal area are equidecomposible.

CHAPTER 4

Similar Triangles

4.1 THE THREE THEOREMS

Informally, similar triangles can be thought of as having the same shape but different sizes. If you had a picture of a triangle and you enlarged it, the result would be similar to the original triangle. Here is the precise definition (Fig. 4.1).

DEFINITION. Two triangles $\triangle ABC$ and $\triangle DEF$ are said to be similar with ratio k if the following six conditions hold for some positive real number k:

1. $\angle A \cong \angle D$
2. $\angle B \cong \angle E$
3. $\angle C \cong \angle F$
4. $DE = k \cdot AB$
5. $EF = k \cdot BC$
6. $DF = k \cdot AC$

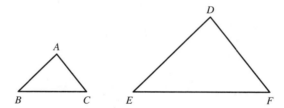

Figure 4.1: $\triangle ABC$ is similar to $\triangle DEF$ with ratio 2

The theory of similar triangles is similar in many ways to the theory of congruent triangles. In particular, we have analogues of the three basic congruence theorems ASA, SAS, and SSS. These similarity theorems are the main subject of this section.

THEOREM(ASA FOR SIMILAR TRIANGLES). If $\angle A \cong \angle D$, $\angle B \cong \angle E$, and if $DE = k \cdot AB$, then $\triangle ABC$ and $\triangle DEF$ are similar with ratio k.

Proof. The proof will be based on comparing the areas of the two triangles. Since $\angle A \cong \angle D$ we know from Section 3.3 that the ratio of the areas is the same as the ratio of the products of the including sides. Hence

$$\frac{\text{area}\triangle DEF}{\text{area}\triangle ABC} = \frac{DE \cdot DF}{AB \cdot AC} = k \cdot \frac{DF}{AC} \tag{4.1}$$

since $DE = k \cdot AB$. Likewise, since $\angle B \cong \angle E$ we get

$$\frac{\text{area}\triangle DEF}{\text{area}\triangle ABC} = \frac{DE \cdot EF}{AB \cdot BC} = k \cdot \frac{EF}{BC}. \tag{4.2}$$

We also know that $\angle C \cong \angle F$, since $\angle A \cong \angle D$, $\angle B \cong \angle E$, and the sum of three angles in each triangle is $180°$. Hence

$$\frac{\text{area}\triangle DEF}{\text{area}\triangle ABC} = \frac{DF \cdot EF}{AC \cdot BC}. \tag{4.3}$$

We now compare these three equations. Comparing (4.1) and (4.3) yields

$$k \cdot \frac{DF}{AC} = \frac{DF \cdot EF}{AC \cdot BC}.$$

We can cancel $\frac{DF}{AC}$ from both sides and then multiply both sides by BC. The result is $EF = k \cdot BC$. Likewise, if we work with (4.2) and (4.3), we obtain $DF = k \cdot AC$. This completes the proof. $\qquad\square$

The other two basic similarity theorems follow from this one in combination with the appropriate theorems for congruent triangles.

THEOREM(SAS FOR SIMILAR TRIANGLES). If $\angle A \cong \angle D$, $DE = k \cdot AB$ and $DF = k \cdot AC$, then $\triangle ABC$ and $\triangle DEF$ are similar with ratio k.

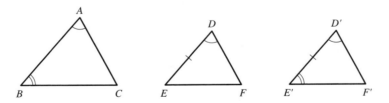

Figure 4.2: Proof of SAS for similar triangles

Proof. We construct a third triangle $\triangle D'E'F'$ such that $\angle D' \cong \angle A$, $\angle E' \cong \angle B$ and $\overline{D'E'} \cong \overline{DE}$. Since $DE = k \cdot AB$, it follows that $D'E' = k \cdot AB$. Thus, by ASA for similar triangles, $\triangle ABC$ and $\triangle D'E'F'$ are similar with ratio k. Hence $D'F' = k \cdot AC$. But DF is also equal to $k \cdot AC$, which implies $D'F' = DF$. Now compare $\triangle DEF$ and $\triangle D'E'F'$: $\angle D \cong \angle D'$, $\overline{DE} \cong \overline{D'E'}$, and $\overline{DF} \cong \overline{D'F'}$ so these triangles are congruent by SAS. Hence $\angle E \cong \angle E' \cong \angle B$ and the theorem now follows by ASA for similar triangles. $\qquad\square$

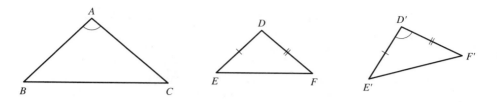

Figure 4.3: Proof of SSS for similar triangles

THEOREM(SSS FOR SIMILAR TRIANGLES). If $DE = k \cdot AB$, $EF = k \cdot BC$, and $DF = k \cdot AC$, then $\triangle ABC$ and $\triangle DEF$ are similar with ratio k.

Proof. Construct a third triangle $\triangle D'E'F'$ such that $\angle D' \cong \angle A$, $\overline{DE} \cong \overline{D'E'}$ and $\overline{DF} \cong \overline{D'F'}$. The proof is now parallel to the previous proof: $\triangle ABC$ and $\triangle D'E'F'$ are similar with ratio k by SAS for similar triangles. This in turn implies that $\overline{EF} \cong \overline{E'F'}$ and consequently $\triangle DEF \cong \triangle D'E'F'$ by SSS. We leave the remaining details as an exercise. □

In most applications of similar triangles we will use a weaker property than the one we have been studying. We say that $\triangle ABC$ and $\triangle DEF$ are similar—without specifying k—if for some k the triangles are similar with ratio k. If you look at the definition at the beginning of this chapter, you will see that k can be calculated from the triangles themselves as

$$k = \frac{DE}{AB}, \text{ or } k = \frac{EF}{BC}, \text{ or } k = \frac{DF}{AC}.$$

This leads to the following definition.

DEFINITION. Two triangles $\triangle ABC$ and $\triangle DEF$ are said to be similar if $\angle A \cong \angle D$, $\angle B \cong \angle E$, $\angle C \cong \angle F$ and if the three numbers $\frac{DE}{AB}$, $\frac{EF}{BC}$, and $\frac{DF}{AC}$ are all equal; that is,

$$\frac{DE}{AB} = \frac{EF}{BC} = \frac{DF}{AC}$$

In this case we write $\triangle ABC \sim \triangle DEF$.

We may now rewrite our three basic theorems using this new definition, without mention of the ratio k.

THEOREM(ASA FOR SIMILAR TRIANGLES). (also called the AA theorem) If $\angle A \cong \angle D$ and $\angle B \cong \angle E$ then $\triangle ABC \sim \triangle DEF$.

THEOREM(SAS FOR SIMILAR TRIANGLES). If $\angle A \cong \angle D$ and $\frac{DE}{AB} = \frac{DF}{AC}$, then $\triangle ABC \sim \triangle DEF$.

THEOREM(SSS FOR SIMILAR TRIANGLES). If $\frac{DE}{AB} = \frac{DF}{AC} = \frac{EF}{BC}$, then $\triangle ABC \sim \triangle DEF$,

Consider an example. Let $AB = 6$, $BC = 12$, and $AC = 15$ and let $DE = 60$, $EF = 120$, and $DF = 150$. Then $\triangle ABC \sim \triangle DEF$ by SSS for similar triangles. We see this by comparing the sides of $\triangle ABC$ with the sides of $\triangle DEF$ and noting that each side of $\triangle DEF$ is 10 times bigger than the corresponding side of $\triangle ABC$ ($k = 10$). However, we could also compare the sides of $\triangle ABC$ to each other and the sides of $\triangle DEF$ to each other and compare the ratios: BC is twice as long as AB and EF is twice as long as DE; AC is $2\frac{1}{2}$ times as long as AB, etc. Algebraically, of course, it is obvious that the equation $\frac{DE}{AB} = \frac{DF}{AC}$ is equivalent to the equation $\frac{DE}{DF} = \frac{AB}{AC}$, and likewise for the other ratios.

Hence we may rewrite the last two theorems.

THEOREM(SAS FOR SIMILAR TRIANGLES). If $\angle A \cong \angle D$ and $\frac{AB}{AC} = \frac{DE}{DF}$, then $\triangle ABC \sim \triangle DEF$.

THEOREM(SSS FOR SIMILAR TRIANGLES). If $\frac{AB}{AC} = \frac{DE}{DF}$ and $\frac{AB}{BC} = \frac{DE}{EF}$ then $\triangle ABC \sim \triangle DEF$.

4.2 APPLICATIONS TO CONSTRUCTIONS

In this section we construct geometric solutions to three types of algebraic equations. The first is $\frac{x}{a} = \frac{b}{c}$, the second type is the pair of simultaneous equations

$$x + y = a$$
$$\frac{x}{y} = \frac{b}{c},$$

and the third type is similar to the second with $x - y$ replacing $x + y$.

PROBLEM. Given line segments of lengths a, b, and c, construct a line segment of length x such that $\frac{x}{a} = \frac{b}{c}$.

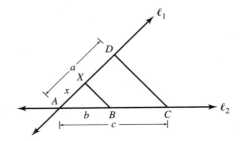

Figure 4.4: Solution of $\frac{x}{a} = \frac{b}{c}$

Solution. Let ℓ_1 be any line. As in Fig. 4.4, construct points A, B, C on ℓ_1 such that $AB = b$ and $AC = c$. Let ℓ_2 be another line through A and construct D such that $AD = a$. Draw $\overline{CD}$ and construct a line through B, parallel to $\overleftrightarrow{CD}$, which intersects ℓ_2 at the point X. Then $AX = x$ is the required distance.

Proof. Since $\overleftrightarrow{CD} \parallel \overleftrightarrow{BX}$, $\angle BCD \cong \angle ABX$ since they are corresponding angles, and $\angle A \cong \angle A$. Hence, by the AA theorem, $\triangle ABX \sim \triangle ACD$. So $\frac{AX}{AD} = \frac{AB}{AC}$, or $\frac{x}{a} = \frac{b}{c}$. □

PROBLEM. Given a line segment $\overline{AB}$ of length a and two other segments with lengths b and c, find a point C in $\overline{AB}$ such that $\frac{AC}{CB} = \frac{b}{c}$.

Solution. Construct a line through A (not equal to $\overleftrightarrow{AB}$) and a line parallel to it through B. Construct points D on the first line and E on the second, as in Fig. 4.5, such that $AD = b$, $BE = c$, and such that D and E are on opposite sides of $\overleftrightarrow{AB}$. Then C will be the intersection point of $\overline{DE}$ and $\overline{AB}$. Moreover, if $AC = x$ and $CB = y$, then x and y are solutions to the system of equations $x + y = a$, $\frac{x}{y} = \frac{b}{c}$.

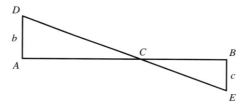

Figure 4.5: Dividing $\overline{AB}$ internally in a given ratio

Proof. Since $\overleftrightarrow{AD} \parallel \overleftrightarrow{BE}$, $\angle D \cong \angle E$ as alternate interior angles. Also, $\angle ACD \cong \angle BCE$ by the vertical angle theorem. Hence $\triangle ACD \sim \triangle BCE$ by AA. So $\frac{AC}{BC} = \frac{AD}{BE} = \frac{b}{c}$. □

PROBLEM. Given a line segment $\overline{AB}$ and two other segments with lengths b and c, with $b > c$, find a point C on $\overleftrightarrow{AB}$ such that $\frac{AC}{BC} = \frac{b}{c}$.

Solution. As in the previous problem, we construct points D and E such that $\overleftrightarrow{AD} \parallel \overleftrightarrow{BE}$, $AD = b$, $BE = c$, but in this case we take D and E to be on the same side of $\overleftrightarrow{AB}$ (Fig. 4.6) .

Proof. As before, the proof follows from $\triangle ACD \sim \triangle BCE$. We omit the details. □

As a further application of these ideas we show how to multiply a line segment by a rational number. For example, say we would like to find a line segment $\overline{AC}$ two-fifths as long as a given segment $\overline{AB}$. We draw another line through A, as in Fig. 4.7, and choose any point A_1 on that line. Then, moving further along the line we construct A_2, A_3, A_4, and A_5 such that $\overline{AA_1} \cong \overline{A_1A_2} \cong \overline{A_2A_3} \cong \overline{A_3A_4} \cong \overline{A_4A_5}$. Join points A_5

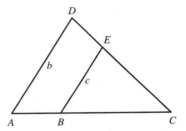

Figure 4.6: Dividing $\overline{AB}$ externally in a given ratio

and B and draw a line ℓ through A_2 parallel to $\overline{A_5B}$. Then C will be the intersection point of ℓ with $\overline{AB}$. By our first construction, $AC = \frac{2}{5} \cdot$ AB.

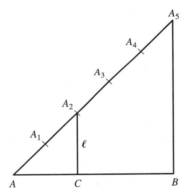

Figure 4.7: Finding AC as two-fifths of AB

PROBLEMS

4.1. Assume that $\triangle ABC$ and $\triangle EFG$ are similar, with ratio k. Let $\overline{AD}$ be an altitude of $\triangle ABC$ and $\overline{EH}$ be an altitude of $\triangle EFG$. Prove that $EH = k \cdot$ AD. What can you conclude about the areas of the two triangles?

*4.2. In triangles $\triangle ABC$ and $\triangle DEF$ assume that $ED = k \cdot$ AB, $FE = k \cdot BC$ and that $\angle C$ and $\angle F$ are right angles. Prove that $\triangle ABC$ and $\triangle DEF$ are similar with ratio k. (This is an analogue of SSA for right triangles.)

*4.3. Let the two lines $\overleftrightarrow{AB}$ and $\overleftrightarrow{CD}$ be cut by the transversals $\overleftrightarrow{AC}$ and $\overleftrightarrow{BD}$, which intersect in the point E. Show that the following are equivalent.

(a) $\overleftrightarrow{AB} \parallel \overleftrightarrow{CD}$

(b) $\triangle EAB \sim \triangle ECD$

(c) $\frac{EA}{EC} = \frac{EB}{ED}$

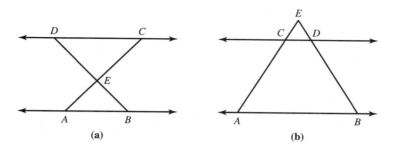

Figure 4.8: Exercise 3 (two cases)

(d)$\frac{EA}{AC} = \frac{EB}{BD}$

4.4. Let $\triangle ABC$ be a right triangle with right angle at C and with altitude $\overline{CD}$. Prove that $\triangle ABC \sim \triangle ACD \sim \triangle CBD$. Use this to give (yet another) proof of the Pythagorean theorem.

4.5. Let $\triangle ABC$ and $\triangle DEF$ be such that $\overleftrightarrow{AB} \parallel \overleftrightarrow{DE}$, $\overleftrightarrow{BC} \parallel \overleftrightarrow{EF}$ and $\overleftrightarrow{AC} \parallel \overleftrightarrow{DF}$. Prove that $\triangle ABC \sim \triangle DEF$. (This is a handy theorem if you need to draw a pair of similar triangles at the blackboard.)

4.6. Assume that you are given a line segment $\overline{AB}$ of length 1 (a unit).
   (a) Given line segments of lengths a and b, construct a segment of length $\frac{a}{b}$.
   (b) Given line segments of lengths a and b, construct a line segment of length ab.
   (c) Given line segments of lengths a, b, and c, construct a line segment of length abc.

4.7. Given line segments of lengths a, b, and c, construct a line segment of length x such that
   (a) $\frac{x^2}{a^2} = \frac{b}{c}$
   (b) (b) $\frac{x}{a} = \frac{b^2}{c^2}$.

4.8. Given a triangle $\triangle ABC$ and a point P on $\overline{AB}$ with $AP > PB$, show how to construct a point Q on $\overline{AC}$ such that $\triangle APQ$ will have one-half the area of $\triangle ABC$.

*4.9. Let ℓ be a line, P a point not on ℓ, and k a positive real number. For each A on ℓ, let B be the point on $\overline{AP}$ such that $BP = k \cdot AP$. Prove that the set S in the plane of all such points B is a straight line parallel to ℓ.

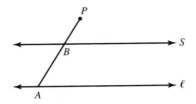

Figure 4.9: Exercise 8

CHAPTER SUMMARY

- $\triangle ABC \sim \triangle DEF$ with ratio k if these conditions hold for a positive real number k:

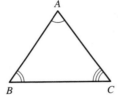

  1. $\angle A \cong \angle D$
  2. $\angle B \cong \angle E$
  3. $\angle C \cong \angle F$
  4. $DE = k \cdot AB$
  5. $EF = k \cdot BC$
  6. $DF = k \cdot AC$

- There are three theorems for proving two triangles similar analogous to the triangle conguence theorems

    THEOREM(ASA FOR SIMILAR TRIANGLES). If $\angle A \cong \angle D$, $\angle B \cong \angle E$ and $DE = k \cdot AB$, then $\triangle ABC \sim \triangle DEF$ with ratio k.

    THEOREM(SAS FOR SIMILAR TRIANGLES). If $\angle A \cong \angle D$, $DE = k \cdot AB$, and $DF = k \cdot AC$, then $\triangle ABC \sim \triangle DEF$ with ratio k.

    THEOREM(SSS FOR SIMILAR TRIANGLES). If $DE = k \cdot AB$, $EF = k \cdot BC$, and $DF = k \cdot AC$, then $\triangle ABC \sim \triangle DEF$ with ratio k.

- Or, using the definition that two triangles are similar if there is some k such that they are similar with ratio k, then the three theorems above can be rewritten as:

    THEOREM(ASA FOR SIMILAR TRIANGLES). (also known as AA): If $\angle A \cong \angle D$ and $\angle B \cong \angle E$, then $\triangle ABC \sim \triangle DEF$.

THEOREM(SAS FOR SIMILAR TRIANGLES). If $\angle A \cong \angle D$ and $\frac{DE}{AB} = \frac{DF}{AC}$, then $\triangle ABC \sim \triangle DEF$.

THEOREM(SSS FOR SIMILAR TRIANGLES). If $\frac{DE}{AB} = \frac{EF}{BC} = \frac{DF}{AC}$, then $\triangle ABC \sim \triangle DEF$.

- Given line segments of length $a, b,$ and c, one can construct a segment with length x such that $\frac{x}{a} = \frac{b}{c}$. On the diagram, ℓ_2 is any line, A, B and C are constructed on ℓ_2 such that $AB = b$ and $AC = c$. ℓ_1 is a line through A with $AD = a$. After $\overline{CD}$ is constructed, x is found by drawing a line at B to X on ℓ_1 parallel to $\overline{CD}$. Thus, using similar triangles, $\frac{x}{a} = \frac{b}{c}$.

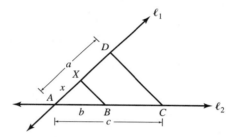

- Given a line segment $\overline{AB}$ and two other segments with lengths b and c, you can construct a point C in $\overleftrightarrow{AB}$ such that $\frac{AC}{CB} = \frac{b}{c}$. On the diagram, $\overline{AD}$ is constructed through A and $\overline{BE}$ through B such that $\overline{AD} \| \overline{BE}$, $AD = b$, and $BE = c$. (D and E are on opposite sides of $\overleftrightarrow{AB}$. Thus, $\frac{AC}{CB} = \frac{b}{c}$.

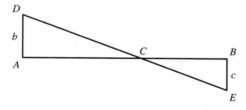

- When $b > c$, the same construction can be made if D and E are constructed such that $\overleftrightarrow{AD} \| \overleftrightarrow{BE}$, $AD = b$, $BE = c$, and D and E are on the same side of $\overleftrightarrow{AB}$, as drawn in the following figure.

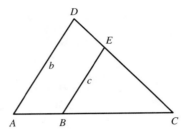

- A line segment can be multiplied by a rational number as well. Starting out with

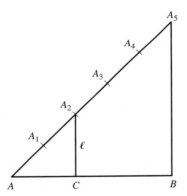

segment $\overline{AB}$ for example, one can find AC as $\frac{2}{5}AB$. Draw a ray up from A at a convenient angle, dividing the ray into 5 equal segments demarcated by A_1 to A_5, connecting A_5 to B, then draw a parallel line from A_2 to C as shown. Then, $AC = \frac{2}{5}AB$.

CHAPTER 5

Circles

5.1 CIRCLES AND TANGENTS

A circle is defined by a center and a radius. Given a point O and a length r, the circle with radius r and center O is defined to be the set of all points A such that $OA = r$. Much of this chapter will be concerned with the relationships among circles and various lines. The lines commonly associated with circles are as follows:

A *secant* is a line that intersects a circle at two points.

A *tangent* is a line that intersects a circle in one point.

A *radius* is a line segment with one end point at the center and the other on the circle. Note, as before, that the word "radius" sometimes refers to the length of the segment.

A *chord* is a line segment with both end points on the circle.

A *diameter* is a chord that contains the center.

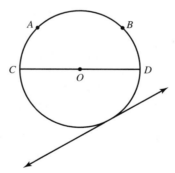

Figure 5.1: Lines and circles

In Fig. 5.1, $\overleftrightarrow{AB}$ and $\overleftrightarrow{CD}$ are secants, ℓ is a tangent, $\overline{OC}$ is a radius, $\overline{AB}$ and $\overline{CD}$ are chords and $\overline{CD}$ is a diameter. We first study intersections of circles and lines.

LEMMA. A circle and a line cannot intersect in three or more points.

Proof. The proof will be by contradiction. Assume that we have a circle with center O and radius r, and a line ℓ with distinct points A, B, and C on both the line and circle. By the definition of circle, $\overline{OA}$, $\overline{OB}$, and $\overline{OC}$ all have length r

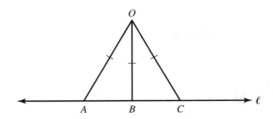

Figure 5.2: Intersection of a circle and a line

and so $\overline{OA} \cong \overline{OB} \cong \overline{OC}$. Hence the triangles $\triangle OAB$, $\triangle OBC$, and $\triangle OAC$ are all isosceles. Hence

$$\angle OAC \cong \angle OCA,$$

$$\angle OAB \cong \angle OBA,$$

and

$$\angle OBC \cong \angle OCB.$$

Comparing the first and third congruences, we see that $\angle OBC \cong \angle OAB$. But this contradicts the exterior angle theorem: Since $\angle OBC$ is an exterior angle of $\triangle OAB$ it must be larger than $\angle OAB$. This contradiction proves the lemma.
□

We see from this lemma that a line and a circle can intersect in either zero, one, or two points. If the line intersects the circle in only one point, it is called a tangent.

THEOREM. Let A be a point on a circle with center O, and let ℓ be a line through A. Then ℓ is tangent to the circle at A if and only if $\overleftrightarrow{OA} \perp \ell$.

Proof. Since this is an "if and only if" theorem, we have two statements to prove: If $\overleftrightarrow{OA}$ is perpendicular to ℓ at A, then ℓ is a tangent, and if ℓ is tangent to the circle at A, then $\overleftrightarrow{OA} \perp \ell$.

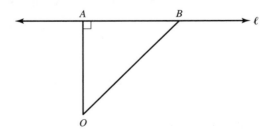

Figure 5.3: A perpendicular to a radius through an end point on the circle is a tangent

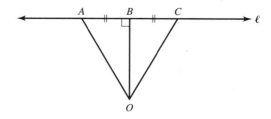

Figure 5.4: A tangent is perpendicular to a radius

We first consider the case in which $\overleftrightarrow{OA}$ is assumed to be perpendicular to ℓ, and we want to show that ℓ is a tangent. So, by way of contradiction we assume that ℓ is not tangent to the circle and that there is another point B on the intersection of the circle with ℓ. Since A and B are both points on the circle, $\overline{OA} \cong \overline{OB}$. But $\triangle OAB$ is a right triangle with hypotenuse $\overline{OB}$. This gives a contradiction because the hypotenuse of a right triangle has to be longer than either of the two legs.

Next, to prove the converse, we assume that $\overleftrightarrow{OA}$ is not perpendicular to ℓ and we will prove that ℓ is not tangent to the circle. In order to do this we will find another point in the intersection of the circle with ℓ.

Since $\overleftrightarrow{OA}$ is not perpendicular to ℓ, we construct B on ℓ such that $\overleftrightarrow{OB}$ is perpendicular. Then we construct C on ℓ (Fig. 5.4) such that $\overline{AB} \cong \overline{BC}$. It is not hard to see that $\triangle OBA \cong \triangle OBC$ by SAS. Thus, $\overline{OA} \cong \overline{OC}$. But since $\overline{OA}$ is a radius, this implies that C is also a point on the circle and so completes the proof.   □

This theorem has a number of consequences. First, we can use it to construct tangents.

PROBLEM. Give a circle C with center O and point A on C, construct a line ℓ through A and tangent to C.

Solution. Line ℓ will be the line perpendicular to $\overleftrightarrow{OA}$ at A.

COROLLARY. If C is any circle and A is a point on C, then there exists a unique line through A tangent to C.

Both the construction and corollary are easy consequences of the theorem and we omit the proofs. We close this section with one more result on tangents.

THEOREM. Let C be a circle with center O and let A and B be points on C. Assume P is a point external to C and that $\overleftrightarrow{PA}$ and $\overleftrightarrow{PB}$ are tangent to C. Then $\overline{PA} \cong \overline{PB}$.

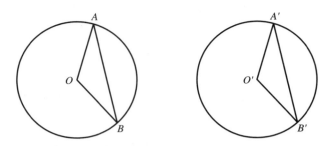

Figure 5.5: Equal angles and equal chords

Proof. Consider the triangles $\triangle OAP$ and $\triangle OBP$. Each has $\overline{OP}$ for one side, $\overline{OA} \cong \overline{OB}$ by definition of a circle, and $\angle A$ and $\angle B$ are each right angles. Hence, $\triangle OAP \cong \triangle OBP$ by SSA for right triangles. So $\overline{PA} \cong \overline{PB}$, as claimed. $\qquad\square$

5.2 ARCS AND ANGLES

In this section we discuss circular arcs and how they are measured. We then prove a theorem relating sizes of arcs to angles inscribed in them. This will prove to be a key result for the rest of our study of circles and also for later chapters. We begin with an easy lemma.

LEMMA. Let C and C' be circles with centers O and O' and with equal radii. Let A and B be points on C and A' and B' be points on C'. Then $\overline{AB} \cong \overline{A'B'}$ if and only if $\angle AOB \cong \angle A'O'B'$.

Proof. (Fig. 5.5) Assume $\overline{AB} \cong \overline{A'B'}$, since C and C' have equal radii, $\overline{OA} \cong \overline{OA'}$ and $\overline{OB} \cong \overline{OB'}$, then $\triangle AOB \cong \triangle A'O'B'$ by SSS, and $\angle AOB$ and $\angle A'O'B'$ are corresponding parts of congruent triangles. Conversely, if we assume that $\angle AOB \cong \angle A'O'B'$, then $\triangle AOB \cong \triangle A'O'B'$ by SAS, and here $\overline{AB} \cong \overline{A'B'}$ are corresponding parts of congruent triangles. $\qquad\square$

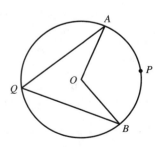

Figure 5.6: Arcs

Now let C be any circle and A and B be two points on C. A and B divide the circle into two arcs, and, confusingly enough, both arcs are referred to as $\overset{\frown}{AB}$. If itis not clear from the context which arc is meant, it is proper to add a point in between, such as $\overset{\frown}{APB}$ or $\overset{\frown}{AQB}$. Assume, as in Fig. 5.6, that $\overset{\frown}{APB}$ is the smaller of the two. Then we will define the measure of the arc $\overset{\frown}{APB}$ to be that of $\angle AOB$ and the measure $\overset{\frown}{AQB}$ to be $360° - \angle AOB$. In the spirit of the lemma we will say that two arcs $\overset{\frown}{AB}$ and $\overset{\frown}{CD}$ are congruent, written $\overset{\frown}{AB} \cong \overset{\frown}{CD}$, if they have the same measure and if they come from circles of equal radii (or from the same circle). So $\overset{\frown}{AB} \cong \overset{\frown}{CD}$ if and only if they have the same measure and if the segments $\overline{AB}$ and $\overline{CD}$ are congruent. Moreover, in the spirit of Chapter 0, we will often identify an arc with its degree measure. Given three points on a circle, such as A, Q, and B in Fig. 5.6, we will say that $\angle AQB$ is *inscribed* in the circle and that it *subtends* arc $\overset{\frown}{AB}$, to refer to $\overset{\frown}{AB}$, the arc $\overset{\frown}{APB}$ that does not contain the vertex Q.

THEOREM. If we are given a circle with center O and containing points A, B, and C so that $\angle ABC$ subtends the arc $\overset{\frown}{AC}$, then

$$\angle B = \frac{1}{2} \overset{\frown}{AC} = \frac{1}{2}\angle AOC.$$

(Notice that in the statement of this theorem we have used our notation convention, which identifies angles and arcs with the real numbers that are their degree measures.)

Proof. There are three possible cases to consider: Either O lies on a side of $\angle ABC$, or it lies on the interior of $\angle ABC$, or it lies outside of $\angle ABC$.

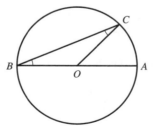

Figure 5.7: $\angle B = \frac{1}{2} \overset{\frown}{AC}$, case of O on $\overline{AB}$

First, let O be on a side of $\angle ABC$: say O is on $\overline{AB}$. Consider the triangle $\triangle OBC$. Since it is isosceles, $\angle B \cong \angle C$. Now

$$180° = \angle B + \angle C + \angle COB$$
$$= 2\angle B + \angle COB.$$

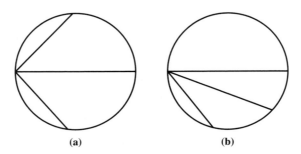

Figure 5.8: $\angle B = \frac{1}{2}\, \overset{\frown}{AC}$, case of O not on $\overline{AB}$

Hence $2\angle B = 180° - \angle COB = \angle COA$. Therefore, $\angle B = \frac{1}{2}\angle AOC = \frac{1}{2}\, \overset{\frown}{AC}$, as claimed.

Next, assume O is inside of $\angle ABC$ as in Fig. 5.8(a). Connect B to O and extend to a point P on the circle so that $\overline{BP}$ will be a diameter. We may now apply the previous case to the angles $\angle ABP$ and $\angle PBC$, since O is on one side of each of them. Hence

$$\angle ABP = \frac{1}{2}\, \overset{\frown}{AP} \ \text{ and } \angle PBC = \frac{1}{2}\, \overset{\frown}{PC}\ .$$

Adding these two equations yields

$$\angle ABP + \angle PBC = \frac{1}{2}(\overset{\frown}{AP} + \overset{\frown}{PC}).$$

The left hand-side of this equation is $\angle ABC$ and the right-hand side is $\frac{1}{2}\, \overset{\frown}{AC}$.

The last case, in which O is outside [see Fig. 5.8(b)], is similar. The only change is that now $\angle ABC = \angle ABP - \angle PBC$.   □

Here are some easy consequences of our theorem.

COROLLARY. If $\angle BAC$ and $\angle BA'C$ are each inscribed in a circle and if each subtends the same arc $\overset{\frown}{BC}$, then $\angle A \cong \angle A'$.

Proof. $\angle A = \frac{1}{2}\, \overset{\frown}{BC}$ and $\angle A' = \frac{1}{2}\, \overset{\frown}{BC}$.   □

COROLLARY. An angle inscribed in a semicircle must be a right angle.

Proof. In this case the arc subtended is $180°$.   □

A third consequence of our theorem is the following slightly more difficult but extremely useful theorem:

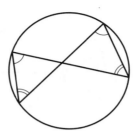

Figure 5.9: $AE \cdot EB = CE \cdot ED$

THEOREM. If we are given a circle with chords $\overline{AB}$ and $\overline{CD}$ which intersect at a point E inside the circle, then AE $\cdot EB = CE \cdot$ ED.

Proof. The angles $\angle A$ and $\angle D$ each subtend the arc $\overset{\frown}{BC}$ and therefore are congruent. Likewise, $\angle C$ and $\angle B$ are congruent, for they each subtend $\overset{\frown}{AD}$. Hence $\triangle ACE \sim \triangle DBE$. So

$$\frac{AE}{DE} = \frac{CE}{BE}.$$

Cross - multiplication now yields the desired result. □

In the situation of the theorem and Fig. 5.9 we can also calculate $\angle AEC$.

THEOREM. Given intersecting chords in a circle (as in Fig. 5.9),

$$\angle AEC = \frac{1}{2}(\overset{\frown}{AC} + \overset{\frown}{BD}).$$

Proof. Since the sum of the angles of $\triangle AEC$ is $180°$, we conclude that

$$\angle AEC = 180° - \angle A - \angle C.$$

Now $\angle A = \frac{1}{2}\overset{\frown}{CB}$ and $\angle C = \frac{1}{2}\overset{\frown}{AD}$. Also, we can express $180°$ in terms of arcs: $180° = \frac{1}{2} \cdot 360° = \frac{1}{2}(\overset{\frown}{AC} + \overset{\frown}{BC} + \overset{\frown}{BD} + \overset{\frown}{AD})$. Substituting all of this into the equation for $\angle AEC$ yields $\angle AEC = 180° - \angle A - \angle C$

$$= \tfrac{1}{2}(\overset{\frown}{AC} + \overset{\frown}{BC} + \overset{\frown}{BD} + \overset{\frown}{AD}) - \tfrac{1}{2}\overset{\frown}{BC} - \tfrac{1}{2}\overset{\frown}{AD}$$

$$= \tfrac{1}{2}(\overset{\frown}{AC} + \overset{\frown}{BD}), \text{ as claimed.} \qquad \square$$

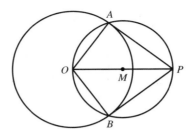

Figure 5.10: Tangent to a circle from an exterior point

5.3   APPLICATIONS TO CONSTRUCTIONS

We now show how these results can be applied to solve some construction problems.

> **PROBLEM.** Given a circle C with center O and a point P outside of C, find a line ℓ that contains P and that is tangent to C.

> **Solution.** Connect $\overline{PO}$ and find the midpoint M. Draw a circle with diameter $\overline{PO}$ by making M the center and $MO = MP$ the radius. This circle will intersect C at two points, A and B. Both $\overleftrightarrow{PA}$ and $\overleftrightarrow{PB}$ are solutions in that both are tangent to C.

> **Proof.** How do we prove that $\overleftrightarrow{PA}$ and $\overleftrightarrow{PB}$ are tangent to C? Recall that, according to a theorem in Section 5.1, we simply need to show that $\overline{OA}$ is perpendicular to $\overleftrightarrow{PA}$ and $\overline{OB}$ is perpendicular to $\overleftrightarrow{PB}$. Now the angle $\angle OAP$ is inscribed in the semicircle $\overset{\frown}{OAP}$ with center M, and the angle $\angle OBP$ is inscribed in the semicircle $\overset{\frown}{OBP}$. Hence, each of them is a right angle.   □

> **PROBLEM.** Given line segments of lengths a and b, construct a line segment of length x so that $x^2 = ab$. In more geometrical terms, construct x such that a square with side x would have area equal to a rectangle with sides of length a and b.

> **Solution.** First, as in Fig. 5.11, construct a line segment $\overline{AB}$ of length $a + b$, with intermediate point E such that $AE = a$ and $EB = b$. Next construct the midpoint M of $\overline{AB}$. Using M as center, we can now draw a circle with center M and with $\overline{AB}$ as diameter. Finally, construct a line perpendicular to $\overleftrightarrow{AB}$ through E. This line will intersect the circle at points C and D. Then $EC = ED = x$.

> **Proof.** Since $\overline{AB}$ and $\overline{CD}$ are chords intersecting at E, $CE \cdot ED = AE \cdot EB = ab$. So all we need to show is that $CE = ED$. Draw $\overline{MC}$ and $\overline{MD}$ and consider the triangles $\triangle MCE$ and $\triangle MDE$. Since $\overline{MC}$ and $\overline{MD}$ are radii, $\overline{MC} \cong \overline{MD}$,

and it is obvious that $\overline{ME} \cong \overline{ME}$. Also, $\angle CEM$ and $\angle DEM$ are right angles. So, by SSA for right triangles, $\triangle MCE \cong \triangle MDE$ and $\overline{CE}$ and $\overline{ED}$ are corresponding parts of congruent triangles.   □

Our third construction is related to the problem of solving a quadratic equation. How would you solve an equation such as $x^2 - 7x + 12 = 0$? You could use the quadratic formula and, in principle, now that we know how to take square roots geometrically, we could try to develop a geometric construction based on the quadratic formula. Another method of solving $x^2 - 7x + 12$ would be to factor it. We write $x^2 - 7x + 12 = (x-?)(x-?)$ and try to find two numbers whose sum is 7 and whose product is 12. Of course, 3 and 4 work and they would be the solutions. The geometric problem we will try to solve is to find two segments with sum a and product equal to a given square b^2. This corresponds to solving the quadratic equation $x^2 - ax + b^2 = 0$.

PROBLEM. Given line segments of lengths a and b, find two lengths whose sum is a and whose product is b^2. (Or, construct a rectangle with a given area and a given semiperimeter.)

Solution. Let $\overline{AB}$ be a line segment of length a, and construct a circle as in Fig. 5.12 with $\overline{AB}$ as diameter. Construct a line ℓ perpendicular to $\overleftrightarrow{AB}$, at point A. Choose a point C on ℓ such that $AC = b$. Draw a line through C, perpendicular to ℓ, and that meets the circle at point D. (See Exercise 5 of this chapter.) Finally, drop a perpendicular from D to $\overline{AB}$, meeting $\overline{AB}$ at E. Then $\overline{AE}$ and $\overline{EB}$ are the solutions to the problem.

Proof. We need to calculate the sum and product of AE and EB. First,

$$AE + EB = AB = a.$$

As for the product, let $\overleftrightarrow{DE}$ intersect the circle at the second point F. Since $\overline{AB}$ and $\overline{DF}$ are intersecting chords, we know that $AE \cdot EB = DE \cdot EF$. To complete the proof we will show that $DE = EF = b$. $\overline{DE}$ is a side of the rectangle $DEAC$, so $DE = AC = b$. To compute EF, let O be the center of

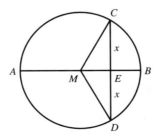

Figure 5.11: Construction of the square root of $AE \cdot EB$

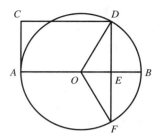

Figure 5.12: Solution of quadratic equation

the circle and draw the radii $\overline{OD}$ and $\overline{OF}$. The triangles $\triangle ODE$ and $\triangle OFE$ are congruent by SSA for right triangles. Hence $FE = DE = b$ and the proof is complete. $\qquad\qquad\qquad\qquad\qquad\qquad\qquad\qquad\qquad\qquad\qquad$ □

5.4   APPLICATION TO QUEEN DIDO'S PROBLEM

Queen Dido was promised as much land as she could cover with an ox hide. In order to get as much as possible, she cut and sewed the hide and made it into a long rope. According to the legend, this is how the ancient city of Carthage was founded.

From a mathematical point of view, here is the problem Queen Dido faced: She wanted to construct a region such that one side would be a straight line (the seashore), the rest of its boundary would be a fixed length (the length of her rope) and its area would be as great as possible. It turns out that the solution to Queen Dido's problem is a region bounded by a semicircle. In this section we will prove this, although we will be a bit vague on a few technical points. For more details about this as well as other interesting geometric optimization problems, refer to Geometric Inequalities, by Nicholas D. Kazarinoff, in the Mathematical Association of America's New Mathematical Library, volume 4, 1961.

There are two ingredients to the proof, which we will prove as separate lemmas.

> **LEMMA 1:.** Of all triangles $\triangle ABC$ with BC equal to a given length a and AC equal to a given length b, the triangle of maximum area is the one with $\angle C = 90°$.

> **LEMMA 2:.** Let $\overline{AB}$ be a fixed segment and let $S = the$ set of all points C (on a given side of $\overleftrightarrow{AB}$) such that $\angle ACB = 90°$. Then S is a semicircle with diameter $\overline{AB}$.

We will present our proofs a bit out of the usual order here. We will assume that the two lemmas are true and use them to prove that the semicircular region is the solution to Queen Dido's problem. Then we will go back and prove the two lemmas.

> ***Proof of Queen Dido's Theorem.***   Let us call the region that maximizes the area R and assume that R touches the seashore along the line segment $\overline{AB}$. Let C be

any point on the rest of the boundary of R. (C is a point of the ox hide.) We first claim that the triangle $\triangle ABC$ is contained in R. The proof of this fact is by contradiction: If the boundary of R sagged in and crossed $\overline{AC}$ or $\overline{BC}$, then by pushing it out we could produce a region with an equal perimeter and greater area and this would contradict our definition of R.

Figure 5.13: Convexity of R

Next, we claim that $\angle ACB$ must be a right angle. Again, this proof will be by contradiction. The triangle $\triangle ABC$ divides R into three regions (Fig. 5.14) we have labeled 1, 2, and 3. If $\angle C \neq 90°$ then we could produce a region with the same perimeter and greater area using lemma 1. If $\angle C < 90°$, we could push A and B further apart to make $\angle C = 90°$. This "pushing" would not affect the lengths of $\overline{AC}$ and $\overline{BC}$, so we could still fit regions 1 and 3 together with our new region 2. In the new figure the area would be greater as guaranteed by lemma 1, and the perimeter would be the same. Since we assumed that R has maximum area, this is impossible and so $\angle C$ is not less than a right angle. We can reason to a similar contradiction if we assumed $\angle C > 90°$. This forces $\angle C = 90°$, as we claimed.

Now we are done, by use of lemma 2. Our region R has a boundary away from shore that consists of points C on a given side of $\overleftrightarrow{AC}$ (the dry side) such that $\angle C = 90°$. Hence the boundary of R is a semicircle.   □

As promised, we omitted some technical details. We would like to point out that one such detail we omitted is that we assumed without proof that the problem has a solution! This means that what we've really proven is that if Queen Dido's problem has a solution, then the solution is given by a semicircle. We now backtrack and provide proofs of our initial lemmas.

Proof of Lemma 1. If $\angle C = 90°$ then $\triangle ABC$ has area $\frac{1}{2}ab$. We will show that if $\angle C \neq 90°$ then $\triangle ABC$ has area less than $\frac{1}{2}ab$. If we take $\overline{BC}$ as the base, then $\triangle ABC$ has area $\frac{1}{2}a \cdot h$, where h is the length of the altitude $\overline{AD}$. But $\overline{AD}$ is a leg of the right triangle $\triangle ADC$, which has hypotenuse $\overline{AC}$. So $AD < AC$, or $h < b$. This implies the area $\frac{1}{2}ah < \frac{1}{2}ab$, as claimed.   □

Proof of Lemma 2. We defined S to be the set $\{C | \triangle ACB = 90°\}$. Let us now define S' to be the semicircle with diameter $\overline{AB}$ and on the appropriate side.

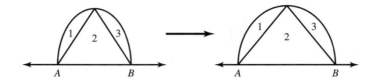

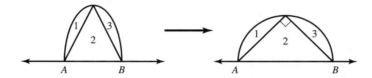

Figure 5.14: Proof that $\angle C = 90°$

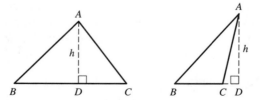

Figure 5.15: Proof of lemma 1

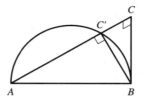

Figure 5.16: Proof of lemma 2

With this notation we need to show that $S = S'$. To show that two sets are equal we need to show first that if C belongs to S', then it belongs to S; then we must show that if C belongs to S, then it belongs to S'.

If C belongs to S' then C is a point on the semicircle with diameter $\overline{AB}$. This means that $\triangle ACB$ is inscribed in a semicircle, and we know that an angle inscribed in a semicircle must be a right angle. So, by definition, C will be an element of S. Conversely, now assume that C belongs to S. This means that we are assuming that $\angle ACB = 90°$ and we want to show that C is on the semicircle with diameter $\overline{AB}$. Our proof will be by contradiction. If C is not on this semicircle, then there is another point C' where $\overleftrightarrow{AC}$ meets the semicircle. As BEFORE, $\angle AC'B$ is a right angle because it is inscribed in a semicircle. Now we can get a contradiction if we consider $\triangle BCC'$. This triangle has two right angles, at C and at C'. This is impossible, and this completes the proof. $\square$

5.5   MORE ON ARCS AND ANGLES

In Section 5.2 we proved various theorems concerning chords. In this section we will prove analogous theorems for secants and tangents.

THEOREM. If B, C, D, and E are points on a circle such that $\overleftrightarrow{BC}$ and $\overleftrightarrow{DE}$ intersect at a point A outside of the circle (as in Fig. 5.17), then

(a) $\angle A = \frac{1}{2}(\overparen{CE} - \overparen{BD})$

(b) $AB \cdot AC = AD \cdot AE$

Proof. (a): Consider $\triangle ACE$. $\angle A = 180° - \angle C - \angle E$. But $\angle C = \frac{1}{2} \overparen{BDE} = \frac{1}{2}(\overparen{BD} + \overparen{DE})$ and $\angle E = \frac{1}{2} \overparen{CBD} = \frac{1}{2}(\overparen{CB} + \overparen{BD})$. Also, $180° = \frac{1}{2} \cdot 360° = \frac{1}{2}[\overparen{CB} + \overparen{BD} + \overparen{DE} + \overparen{CD}]$. Now, by substitution,

$$\angle A = \frac{1}{2}[\overparen{CB} + \overparen{BD} + \overparen{DE} + \overparen{CE}] - \frac{1}{2}[\overparen{BD} + \overparen{DE}] - \frac{1}{2}[\overparen{CB} + \overparen{BD}]$$

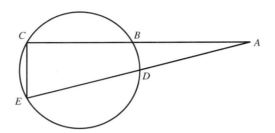

Figure 5.17: Secants intersecting outside of a circle

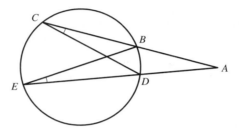

Figure 5.18: Proof that $AB \cdot AC = AD \cdot AE$

$$= \frac{1}{2}[\overset{\frown}{CE} - \overset{\frown}{BD}].$$

(b) For this half of the theorem we consider the triangles $\triangle ACD$ and $\triangle AEB$ (See Fig. 5.18). Each has $\angle A$ for one angle. Also, $\angle C \cong \angle E$ since each subtends the arc $\overset{\frown}{BD}$. Thus, by AA, $\triangle ACD \sim \triangle AEB$. Hence $\frac{AC}{AE} = \frac{AD}{AB}$. If we cross multiply we get (b). $\qquad\qquad\square$

We now turn to the case of tangents. In order to repeat the proof from the secant case, we need this preliminary result.

THEOREM. Assume at a point A on the circle that there is a chord $\overline{AB}$ and a tangent $\overleftrightarrow{AC}$. Then $\angle BAC = \frac{1}{2} \overset{\frown}{AB}$.

We remark that the line segment $\overline{AB}$ and the line $\overleftrightarrow{AC}$ make two different angles with each other (they are supplementary) and that there are two arcs that could be labeled $\overset{\frown}{AB}$. Each angle is half of the arc $\overset{\frown}{AB}$ that it cuts off—the larger angle corresponds to the larger arc and the smaller angle corresponds to the smaller arc.

Proof. Assume that C is such that $\angle BAC$ is less than or equal to $90°$, as in the diagram. Let O be the center of the circle and extend $\overline{AO}$ to a diameter $\overline{AA'}$. Then $\overline{AA'}$ is perpendicular to $\overline{AC}$ and $\angle BAC = 90° - \angle BAA'$. But $\angle BAA' = \frac{1}{2} \overset{\frown}{BA'}$ and $90° = \frac{1}{2} \cdot 180° = \frac{1}{2}(\overset{\frown}{AB} + \overset{\frown}{BA'})$. The result now follows

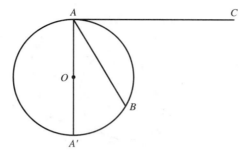

Figure 5.19: $\angle BAC = \frac{1}{2} \overset{\frown}{AB}$

by subtraction. The proof in the case of $\angle BAC$ obtuse is the same, except that $\angle BAC$ will be $90° + \angle BAA'$ rather than $90° - \angle BAA'$.   □

With this tool in hand, we leave it to you to prove the following theorems.

THEOREM. If B, C, and D are points on a circle such that the tangent at B and the secant $\overleftrightarrow{CD}$ intersect at a point A, then

(a) $\angle A = \frac{1}{2}(\overset{\frown}{BD} - \overset{\frown}{BC})$

(b) $AB^2 = AC \cdot AD$

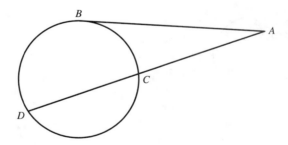

Figure 5.20: A tangent and a secant

THEOREM. Suppose B and C are points on a circle such that the tangent at B and the tangent at C meet at a point A. Let P and Q be points on the circle such that $B\overset{\frown}{Q}C > B\overset{\frown}{P}C$. Then $\angle A = \frac{1}{2}(B\overset{\frown}{Q}C - B\overset{\frown}{P}C)$.

PROBLEMS

5.1. Find the indicated parts in each figure in Fig. 5.22:

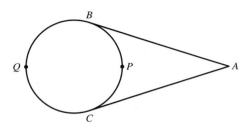

Figure 5.21: Two tangents

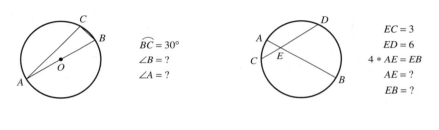

$\widehat{BC} = 30°$
$\angle B = ?$
$\angle A = ?$

$EC = 3$
$ED = 6$
$4 * AE = EB$
$AE = ?$
$EB = ?$

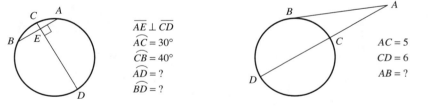

$\overline{AE} \perp \overline{CD}$
$\widehat{AC} = 30°$
$\widehat{CB} = 40°$
$\widehat{AD} = ?$
$\widehat{BD} = ?$

$AC = 5$
$CD = 6$
$AB = ?$

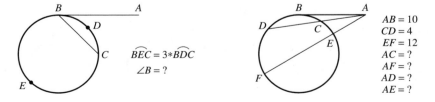

$\widehat{BEC} = 3 * \widehat{BDC}$
$\angle B = ?$

$AB = 10$
$CD = 4$
$EF = 12$
$AC = ?$
$AF = ?$
$AD = ?$
$AE = ?$

Figure 5.22: Exercise 1

5.2. (a) Prove that the perpendicular bisector of a chord of a circle is diameter.

(b) Given a circle, how would you construct the center?

*5.3. Given a quadrilateral $ABCD$ such that each of the four sides is tangent to a circle. ($ABCD$ is said to be circumscribed about the circle), prove that $AB + CD = AD + BC$.

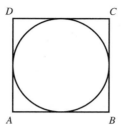

Figure 5.23: Exercise 3

*5.4. Given a circle with center O and a point on the circle A, consider the circle with diameter $\overline{OA}$. Let a line through A intersect the smaller circle at B and the larger circle at C. Prove that $\overline{AB} \cong \overline{BC}$.

5.5. (a) In our solution of the quadratic equation in Section 5.3 we made an assumption. We assumed that the line through C perpendicular to ℓ intersects the circle. What if it doesn't? Find constraints on a and b that ensure that the point D exists.

(b) If the line through C perpendicular to ℓ is a secant, then there are two possible choices for the point D. Are there two different solutions to the problem?

5.6. Use Fig. 5.24 to show how to construct a pair of lengths x, y with given difference $y - x = a$ and given product $xy = b^2$. Make sure to prove that your construction works! (This construction corresponds to solving a quadratic equation such as $x^2 + x - 12$ with one positive and one negative root.)

*5.7. Let C be a circle with center O and radius r. Let P be a point such that $OP = d > r$. Find the length of the tangent line from P to C.

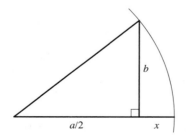

Figure 5.24: Exercise 6

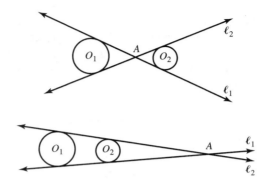

Figure 5.25: Internal and external tangent

5.8. Use figure 5.11 to prove that for any lengths a and b, $\sqrt{ab} \le \frac{1}{2}(a + b)$. This fact is sometimes called the *algebraic-geometric mean inequality*.

5.9. Let $\overline{AB}$ and $\overline{CD}$ be parallel chords in a circle. Prove that $\angle CAB \cong \angle DBA$.

*5.10. Let P be a point outside of a circle with center O, and let $\overline{OP}$ intersect the circle at Q. Prove that Q is the point on the circle closest to P.

5.11. (a) Given concentric circles, prove that two chords of the larger circle that are tangent to the smaller circle are congruent.

(b) Given a circle with center C, radius r, and length d, construct a line through P such that the length of the chord is d.

5.12. Given a circle with center O and three tangent lines, ℓ_1, ℓ_2, and ℓ_3 such that ℓ_1 and ℓ_2 are parallel and ℓ_3 intersects ℓ_1 at A and ℓ_2 at B, prove that $\angle AOB = 90°$.

The next exercises are concerned with the problem of finding a line that will be tangent to two given circles. As Fig. 5.25 illustrates, there are two types of such tangents—internal tangents and external tangents.

5.13. Let circles C_1 and C_2 have centers O_1 and O_2. Let ℓ_1 and ℓ_2 both be externally tangent (or both internally tangent) to C_1 and C_2. Let ℓ_1 and ℓ_2 intersect at A. Prove that A, O_1, and O_2 lie on a straight line. [In Fig. 5.21, prove that the bisector of $\angle A$ contains the center of the circle].

5.14. In Exercise 13, assume that the circles have radii r_1 and r_2, with $r_2 > r_1$. Prove that

$$\frac{AO_1}{AO_2} = \frac{r_1}{r_2}.$$

Use this to give a construction of external tangents to two circles. Prove a corresponding formula and construction for internal tangents.

5.15. Use Fig. 5.26 to give another method to construct a common external tangent to two circles. [Hint: You can construct O_1' since $\overline{O_1'O_2} \cong \overline{O_1O_2}$ and $O_1O_1' = 2r_2 - 2r_1$, and you can use O_1' to find P.] Develop a similar construction for internal tangents.

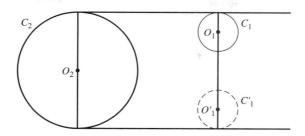

Figure 5.26: Exercise 15

5.16. Given a line segment $\overline{AB}$, define S to be the set of points C on a given side of $\overleftrightarrow{AB}$ such that $\angle ACB = 45°$. What can you say about S?

5.17. If Queen Dido's rope was 2 miles long, how much land did she get?

5.18. Use the solution of Queen Dido's problem to prove the following: Of all the regions with a given perimeter, the circle has the greatest area. [Hint: Let A and B be points on the boundary which divide the perimeter in half.]

5.19. What if instead of an ox hide Queen Dido was given a stick that she was allowed to break into four pieces to form a quadrilateral?

   (a) Assume that quadrilateral $ABCD$ has perimeter p and that it has area greater than or equal to all other quadrilaterals of perimeter p. Use lemma 1 of section 5.4 to prove that $ABCD$ must be a rectangle.

   (b) Prove that of all rectangles of perimeter p the square of side $p/4$ has the greatest area. (One way to do this exercise would be to use Exercise 8.)

*5.20. Continuing in the vein of the previous exercise, if $ABCDEF$ is a hexagon of perimeter p and if it has area greater than or equal to all other hexagons of perimeter p, then it can be inscribed in a circle.

5.21. **(a)** Prove that of all triangles $\triangle ABC$ with $AB = c$ a fixed length and with $BC + AC = a + b = 2m$ a fixed length, the one with $a = b = m$ has the largest area.

   (b) Use part (a) to prove that of all triangles with a given perimeter, the equilateral triangle has the largest area.

   (c) Use part (a) to prove that if $ABCDEF$ is a hexagon of perimeter p and with maximum possible area, then all sides are equal. Combined with Exercise 16, this proves that $ABCDEF$ is regular. [Hint: If all six sides are not equal, then there is a pair of sides that are not equal.]

CHAPTER SUMMARY

- Radius $\overline{OA}$ is perpendicular to tangent $\overleftrightarrow{AC}$ at the point of tangency, point A.

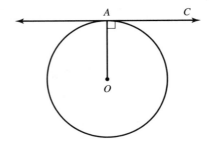

- If $\overline{PA}$ and $\overline{PB}$ are tangent to circle O, then $\overline{PA} \cong \overline{PB}$.

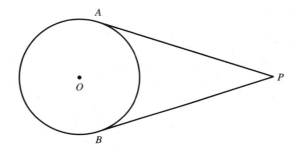

- If $\angle\alpha \cong \angle\alpha'$ and $\overline{OB} \cong \overline{O'B'}$, then $\overset{\frown}{AB} \cong \overset{\frown}{A'B'}$.

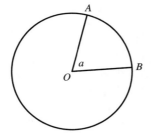   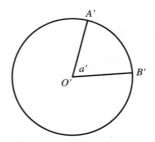

- If two chords meet at a point on the circle, then $b = \angle B = \frac{1}{2}\overset{\frown}{AC}$

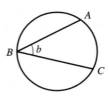

- If two chords intersect inside the circle, as shown, then $\alpha = \frac{1}{2}(\overset{\frown}{AB} + \overset{\frown}{CD})$, $\angle A = \angle A'$, and $AE \cdot ED = BE \cdot EC$.

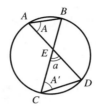

- Given two intersecting secants, as shown, then $\alpha = \frac{1}{2}(\overset{\frown}{AC} - \overset{\frown}{BD})$ and $AE \cdot EB = CE \cdot ED$

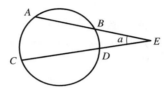

- Given an intersecting chord and tangent, as shown, $\alpha = \frac{1}{2}(\overset{\frown}{AC} - \overset{\frown}{AD})$ and $AB^2 = BD \cdot BC$. This can be seen to be a limiting case of the previous formulas.

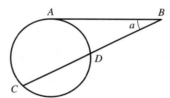

- If a tangent and a chord intersect at a point on the circle, as shown, then $\alpha = \frac{1}{2}\overset{\frown}{AC}$.

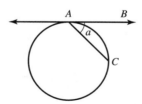

- Given two intersecting tangents, as shown, then $\alpha = \frac{1}{2}(\overset{\frown}{ADB} - \overset{\frown}{AB})$.

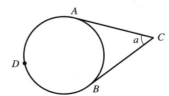

- If $\overset{\frown}{BAC}$ is a semicircle, then $\angle A = 90°$. The converse is also true.

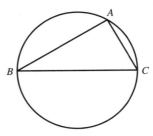

- Construction of tangents from exterior point P: M is the center of the circle with diameter $\overline{OP}$. A and B are intersection points of the two circles. $\overline{PA}$ and $\overline{PB}$ are the desired tangents.

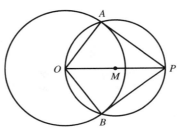

- Construction of the square root of $AE \cdot EB$: Let $AE = a$ and $EB = b$. M is the midpoint of $\overline{AB}$, and $CE = ED = x$. Then $ab = x^2$ as desired.

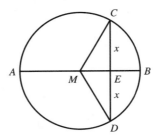

- Solution of Quadratic Equation: (Given segments a and b, find two lengths whose sum equals a and whose product is b^2.) Let $\overline{AB} = a$ and construct circle O with diameter $\overline{AB}$. In the diagram, $\overline{AC} \perp \overline{AB}$, $AC = b$, $\overline{AC} \perp \overline{CD}$ and $\overline{DE} \perp \overline{CD}$. Then $AE + EB = AB = a$ and $AE \cdot EB = DE^2 = b^2$ as desired.

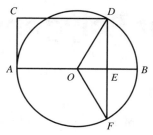

CHAPTER 6

Regular Polygons

6.1 CONSTRUCTIBILITY

This section will be of a character slightly different from the rest of the book. It will contain a number of theorems that we will state and discuss but that we will not prove, because their proofs would involve a lot of extraneous theory, especially from abstract algebra. And, among the theorems we do prove, some of our proofs will resort to hand waving to avoid confusing notation and induction arguments. The material in this chapter will not be used in later chapters, and you may choose to omit it. But we believe that the story we tell is an interesting one and worth including, even at the cost of being less rigorous and complete than usual.

An n-gon (n-sided polygon) is said to be regular if all n sides are congruent to each other and all angles are congruent to each other. Two well-known examples are the square ($n = 4$) and the equilateral triangle ($n = 3$). The first question we will treat is how regular n-gons are constructed.

It is traditional in Euclidean geometry to try to do constructions with an unmarked straightedge and a compass. Most of this section will discuss the regular polygons, which are constructible with this equipment. For the moment, however, let us assume that we have additional equipment available to us—such as a protractor—and that we can draw angles of any size. Using such equipment we can now construct regular n-gons for any $n \geq 3$. For the sake of concreteness, we take $n = 10$ (Fig. 6.1). First

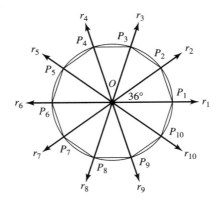

Figure 6.1: Construction of regular polygons

we calculate $\frac{360°}{n}$, which in this case is $36°$. We choose any point O—which will be the center of our regular 10-gon—and draw a ray $\overrightarrow{r_1}$ with initial point O. Next, we draw nine more rays $\overrightarrow{r}_2$, $\overrightarrow{r_3}$, ..., $\overrightarrow{r_{10}}$, each with initial point O and making angles of $36°$, $72°$, $108°$, ...,$324°$ with the ray $\overrightarrow{r_1}$. Now we draw a circle with center O and any radius, which intersects the ten rays in the ten points P_1, P_2,...,P_{10}. Then $P_1 P_2 P_3 \cdots P_{10}$ will be the required 10-gon.

To prove this, you would show that the ten triangles $\triangle O P_1 P_2$, $\triangle O P_2 P_3$, $\triangle O P_3 P_4$, etc. are all congruent to each other by SAS. It would then follow that $\overline{P_1 P_2} \cong \overline{P_2 P_3} \cong \cdots$ and that $\angle P_1 \cong \angle P_2 \cong \cdots$.

It follows from the concepts introduced in this discussion that regular n-gons exist for each $n \geq 3$ and that we can construct them if we can construct angles of size $\frac{360°}{n}$. The converse is also true. So the problem of constructing a regular n-gon with straightedge and compass is equivalent to the problem of constructing a $\frac{360°}{n}$ angle with straightedge and compass. Stated formally, we have sketched a proof of the following theorem.

THEOREM. For a given integer $n \geq 3$, it is possible to construct a regular n-gon if and only if it is possible to construct an angle of size $\frac{360°}{n}$.

QUESTION. For which values of n is it possible to construct a regular n-gon?

We will call a regular n-gon constructible if we can construct it with straightedge and compass. Our theorem allows us to get some partial results right away.

COROLLARY 1. Let $n = 2m$, where $m \geq 3$. If the regular m-gon is constructible, then the regular n-gon is also constructible.

Proof. Let $\alpha = \frac{360°}{m}$ and $\beta = \frac{360°}{n}$. If we use our theorem, we see that the statement "a regular m-gon is constructible" is equivalent to the statement "an angle of size α is constructible"; and the statement "a regular n-gon is constructible" is equivalent to the statement "an angle of size β is constructible." Hence, in order to prove our corollary we may prove, instead, the equivalent statement: If the angle α is constructible, then the angle β is constructible. But, $\beta = \frac{360°}{n} = \frac{360°}{2m} = \frac{1}{2} \cdot \alpha$. The corollary is now immediate, because if we can construct α we can certainly construct $\frac{1}{2}\alpha$, since we have a construction for bisecting angles. $\qquad\square$

Corollary 1 has a number of consequences. For example, since we know that a regular 4-gon is constructible, we can conclude that a regular 8-gon is also constructible. Also, by repeated applications of the corollary, a regular 16-gon, a regular 32-gon, etc. would also be constructible. Likewise, since a regular 3-gon is constructible, so is a regular 6-gon, 12-gon, 24-gon, etc. We also see from this that there are infinitely many values of n for which regular n-gons are constructible, a fact that was not previously apparent.

We will see in a moment that the converse of Corollary 1 is also true. This reduces the questions of constructibility of regular n-gons to the case of odd values of n. Why is this? Suppose you are given an even value of n, such as $n = 120$ and you wanted to know if it was possible to construct a regular 120-gon. You could factor 120 into a product of a power of two times an odd number, $120 = 2 \times 2 \times 2 \times 15$. By applying Corollary 1 and its converse three times, we see that a regular 120-gon is constructible if and only if a regular 15-gon is constructible. In a similar manner, for any even n there is an odd m such that the constructibility of a regular n-gon is equivalent to the constructibility of a regular m-gon. Hence, we now turn to the case of odd n. It will turn out that there are only finitely many odd numbers n such that regular n-gons are contructible. Corollary 2 is useful in describing these values of n.

> **COROLLARY 2.** Let $n = mk$ where $m \geq 3$ and k are integers. If a regular n-gon is constructible, then a regular m-gon is constructible.

> ***Proof.*** The proof of Corollary 2 is very similar to the proof of Corollary 1. Let $\alpha = \frac{360°}{n}$ and $\beta = \frac{360°}{m}$. Then Corollary 2 is equivalent to the statement that if an angle of size α is constructible then an angle of size β is constructible. Now $\beta = \frac{360°}{m} = k \cdot \frac{360°}{m \cdot k} = k \cdot \frac{360°}{n} = k \cdot \alpha$. It is immediate that if we can construct an angle of size α, then we can construct any integral multiple of α. □

Be careful here, because the converse of Corollary 2 is false. It is also not true that if a regular m-gon is contructible and if a regular k-gon is constructible, then a regular mk-gon is constructible. For example, a regular 3-gon is certainly constructible, but it is impossible to construct a regular 9-gon with straightedge and compass. This statement will be true if k and m are relatively prime, but the proof involves some number theory and so we hide it away in the exercises.

Before stating the general solution to our problem, we first tell what happens when n is a prime number. In this case the constructible n-gons are the 3-gon, the 5-gon, the 17-gon, the 257-gon and the 65,537-gon. The primes 3, 5, 17, 257 and 65,537 are called Fermat primes, after Pierre Fermat. Note that $3 = 2^1 + 1$, $5 = 2^2 + 1$, $17 = 2^4 + 1$, $257 = 2^8 + 1$, and $65,537 = 2^{16} + 1$; moreover, the exponents are all powers of 2. Fermat conjectured that $2^{(2^n)} + 1$ was always a prime number, for any value of n. It is now known that Fermat was wrong! For example, $2^{32} + 1$ is not prime; it is divisible by 641. It is not currently known whether there are any prime numbers of the form $2^{(2^n)} + 1$ for $n > 4$, and this is an important open question. It is known, however, that the following fact holds.

> **THEOREM.** If n is a prime number, then a regular n-gon is constructible if and only if n is a Fermat prime.

For more general n we know we may bootstrap to a proof of the next theorem.

> **THEOREM.** (a) If n is odd, then a regular n-gon is constructible if and only if n is a product of distinct Fermat primes.
>   (b) In general, a regular n-gon is constructible if and only if n is a power of 2 times a product of distinct Fermat primes.

We conclude this section with a few examples. For various values of n we tell whether or not is it possible to construct regular n-gons.

$n = 75$. The regular 75-gon is not constructible; $75 = 3 \times 5 \times 5$, and, although 3, 5 and 5 are Fermat primes, they are not distinct.

$n = 120$. The regular 120-gon is constructible; $120 = 2 \times 2 \times 2 \times 3 \times 5$, which is a power of 2 ($2 \times 2 \times 2$) times two distinct Fermat primes.

$n = 23$. The regular 23-gon is not constructible; 23 is a prime which is not a Fermat prime.

6.2   IN THE FOOTSTEPS OF ARCHIMEDES

In this section we will use regular polygons to approximate the areas and circumferences of circles. The method we use was known to the ancient Greeks. Archimedes' approximation of π using regular polygons is considered to be one of the masterpieces of ancient mathematics.

Assume we are given a circle of radius r and that we wish to estimate the area A and circumference C. To get lower bounds we will use inscribed polygons. A polygon is said to be inscribed in a circle if every vertex of the polygon lies on the circle or, equivalently, every side of the polygon is a chord of the circle. Given a circle and an inscribed polygon, the polygon must have a smaller area and a smaller perimeter. We will use these facts to get lower bounds for A and C.

Given any circle with center O, it is fairly easy to construct a regular inscribed hexagon (6-gon).

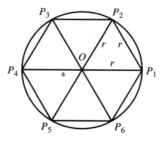

Figure 6.2: Regular inscribed hexagon

Choose any point P_1 on the circle and construct points P_2, P_3, P_4, P_5 and P_6 on the circle such that $P_1 P_2 = r$, ..., $P_5 P_6 = r$. Without giving a detailed proof, the reason this construction works is that the triangles $\triangle O P_1 P_2$, $\triangle O P_2 P_3$, etc. are equilateral triangles. From our observations, the perimeter of the inscribed hexagon gives a lower bound for the circumference of the circle and the area of the hexagon gives a lower bound for the area of the circle. Since each side of the hexagon has length r, the perimeter is $6r$. Hence,

$$6r < C.$$

To calculate the area of the hexagon we will use the fact that the six triangles $\triangle O P_1 P_2$,

$\triangle O P_2 P_3,\ldots,\triangle O P_6 P_1$ are all congruent, so that the area of the hexagon is six times the area of $\triangle O P_1 P_2$. Let $\overline{O D}$ be the altitude, $O D = h$. Since $\triangle O P_1 P_2$ is equilateral,

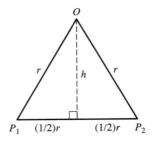

Figure 6.3: Area of $\triangle O P_1 P_2$

D is the midpoint of $\overline{P_1 P_2}$ and we may now use the Pythagorean theorem to calculate h: $h^2 + (\frac{1}{2}r)^2 = r^2$ so $h = \sqrt{3/4}r$. Hence, $\triangle O P_1 P_2$ has area $1/2\sqrt{3/4}r^2$ and the hexagon has area $3\sqrt{3/4}r^2 = (3/2)\sqrt{3}r^2$. Therefore,

$$(3/2)\sqrt{3}r^2 < A.$$

We remark that $(3/2)\sqrt{3} \doteq 2.6$.

We may now improve these crude estimates to obtain whatever accuracy we like by using larger values of n. We will illustrate with the case of $n = 12$. The regular 12-gon (dodecagon) can be constructed using the regular 6-gon. We construct the angle bisectors of each of the six central angles. These lines meet the circle at the points Q_1, Q_2, Q_3, Q_4, Q_5, and Q_6, which bisect the arcs $\overset{\frown}{P_1 P_2}$, $\overset{\frown}{P_2 P_3}$, $\ldots$, $\overset{\frown}{P_6 P_1}$. Then the

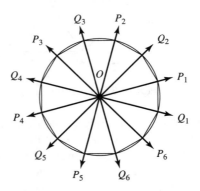

Figure 6.4: Construction of regular 12-gon

polygon $P_1 Q_1 P_2 Q_2 P_3 Q_3 \cdots P_6 Q_6$ will be a regular 12-gon.

The perimeter of this 12-gon will be 12 times the length of the side $\overline{P_1 Q_1}$.

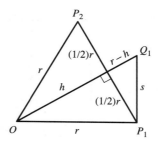

Figure 6.5: Calculation of s

Call this length s. From Fig. 6.5 we see that $s^2 = (\frac{1}{2}r)^2 + (r-h)^2$, by the Pythagorean theorem. And we already calculated that $h = \frac{\sqrt{3}}{2}r$. So

$$s^2 = (\frac{1}{2}r)^2 + (r - \frac{\sqrt{3}}{2}r)^2$$

$$= \frac{1}{4}r^2 + (1 - \frac{\sqrt{3}}{2})^2 r^2$$

$$= \frac{1}{4}r^2 + (\frac{7}{4} - \sqrt{3})r^2$$

$$= (2 - \sqrt{3})r^2.$$

Hence, $s = (\sqrt{2 - \sqrt{3}})r$ and

$$12(\sqrt{2 - \sqrt{3}})r < C.$$

We may calculate this as $(6.21)r < C$.

The area will be 12 times the area of $\triangle OP_1Q_1$. If we take $\overline{OP_1}$ as the base, the height will be $\frac{1}{2}r$. So area $(\triangle OP_1Q_1) = \frac{1}{2}(\frac{1}{2}r)r = \frac{1}{4}r^2$. Multiplying by 12 yields

$$3r^2 < A.$$

Rather than continuing in this vein, we turn to the question of computing upper bounds for C and A. To do this we will use regular polygons circumscribed about the circle. A polygon is said to be circumscribed about the circle if every side of the polygon is tangent to the circle. It is clear in this case that the area of the polygon will be greater then the area of the circle. It is also true that the perimeter of the polygon will be greater than the circumference of the circle. We observe as a technical point that this fact is neither obvious nor easy to prove. We will assume it in order to do our calculations.

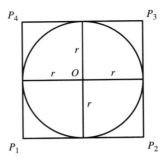

Figure 6.6: Circle with circumscribed square

The easiest polygon to circumscribe about the circle is the square. It has sides equal to $2r$; consequently, it has perimeter $= 8r$ and area $= 4r^2$.

Hence

$$C < 8r \text{ and } A < 4r^2.$$

Using the circumscribed square, we may construct a regular octagon (8-gon) circumscribed about the circle. To do this we draw the diagonals $\overline{P_1P_3}$ and $\overline{P_2P_4}$ of the square. These meet the circle at points A, B, C, D in Fig. 6.7. If we then draw

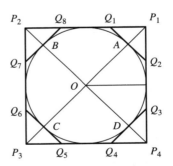

Figure 6.7: Construction of regular octagon

the tangents at these points we get the regular octagon $Q_1Q_2Q_3Q_4Q_5Q_6Q_7Q_8$. We now calculate the perimeter and area of this figure. First, $\overline{OP_1}$ is the hypotenuse of a right triangle, each of whose sides is r. Hence, $OP_1 = \sqrt{2}r$. Since $OA = r$ we get that $AP_1 = \sqrt{2}r - r = (\sqrt{2} - 1)r$. But $AQ_1 = AQ_2 = AP_1$, so the side $Q_1Q_2 = 2(\sqrt{2} - 1)r$. If we multiply by 8 we get

$$C < 16(\sqrt{2} - 1)r,$$

or $C < (6 \cdot 63)r$.

Finally, the area of the octagon equals the area of the square minus four times the area of $\triangle P_1Q_1Q_2$. This triangle has base $2(\sqrt{2} - 1)r$ and height $(\sqrt{2} - 1)r$. Hence the area of the octagon is

$$4r^2 - 4 \cdot \frac{1}{2} \cdot 2(\sqrt{2} - 1)^2 r^2 = 8(\sqrt{2} - 1)r^2,$$

or $A < 3.31\ r^2$.

Combining our results, we see that $12(\sqrt{2 - \sqrt{3}})r < C < 16(\sqrt{2} - 1)r$ or $6.21r < C < 6.63r$, and $3r^2 < A < 8(\sqrt{2} - 1)r^2$ or $3r^2 < A < 3.31r^2$. You likely learned in a previous course that $C = 2\pi r \approx 6.28r$ and $A = \pi r^2 \approx 3.14r^2$. It is not possible to prove these results without saying quite a bit about limits. We do not intend to do so. However, your may observe that the bounds for C were all of the form some constant times r and that by taking n larger we could increase the lower bound and decrease the upper bound. Likewise, we can sandwich A between expressions with constants times r^2.

PROBLEMS

6.1. The regular n-gon we constructed in Section 6.1 is inscribed in a circle. Prove that it can also be circumscribed about a circle.

*6.2. Use the facts that regular triangles and pentagons are constructible to prove that regular 15-gons are constructible.

6.3. Here is a fact from number theory: If n and m are relatively prime (i.e. they have no common factor other than 1) then there exist integers a, b such that $an + bm = 1$. Use this fact to prove that if n and m are relatively prime and if regular n-gons and regular m-gons are constructible, then regular nm-gons are constructible.

6.4. For which n satisfying $3 \leq n \leq 25$ is it possible to construct a regular n-gon?

6.5. For which of these n is a regular n-gon constructible?
(a) 144 (b) 85 (c) 176 (d) 100 (e) 128

6.6. How many odd numbers n are there such that the regular n-gon is constructible? (Assume that there are five Fermat primes.)

6.7. Here is the basic idea of how we construct a regular pentagon: Let $\triangle ABC$ be an isosceles triangle in which $\angle A = 36°$, $\angle C = 72°$. Let $\overline{CD}$ be the bisector of $\angle C$. Then $\overline{AD} \cong \overline{CD} \cong \overline{BC}$, and $\triangle ABC \sim \triangle CBD$. Prove these facts and use them to prove that $BC = (\frac{\sqrt{5}-1}{2})AB$. How could you use this to construct a regular pentagon?

6.8. What lower bounds do you get for C and A if you use a square inscribed in the circle?

6.9. What upper bounds do you get for C and A if you use a hexagon circumscribed about the circle?

6.10. Prove: If $s =$ the length of a side of an n-gon inscribed in a circle of radius r and if $t =$ the length of a side of a $2n$-gon inscribed in a circle of radius r, then

$$t^2 = 2r(r - \sqrt{r^2 - \frac{1}{4}s^2}).$$

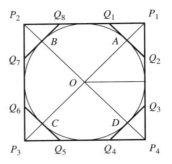

Figure 6.8: Exercise 7

6.11. Use Exercise 10 to get a lower bound for C based on the regular 24-gon; the regular 48-gon.

6.12. Prove that a regular n-gon inscribed in a circle of radius r and with sides of length s has area $\frac{n}{2}s\sqrt{r^2 - \frac{1}{4}s^2}$. Use this formula together with Exercise 10 to get a lower bound for A based on the regular 24-gon and on the regular 48-gon.

CHAPTER SUMMARY

- An n-gon (a polygon with n sides) is regular if all n sides are congruent and all angles are congruent, such as in a square.

- For a given $n \geq 3$, it is possible to construct a regular n-gon if and only if it is possible to construct an angle of size $\frac{360°}{n}$.

- If n is a prime number, a regular n-gon is constructible if and only if n is a Fermat prime. The primes 3, 5, 17, 257, 65,537 are Fermat primes since

$$3 = 2^1 + 1$$
$$5 = 2^2 + 1$$
$$17 = 2^4 + 1$$
$$257 = 2^8 + 1$$

and

$$65,537 = 2^{16} + 1,$$

where all exponents are a power of 2.

- If n is odd, a regular n-gon is constructible if and only if n is a product of Fermat primes. In general, a regular n-gon is constructible if and only if n is a power of 2 times a product of distinct Fermat primes.

- A polygon is inscribed in a circle if every one of its vertices lies on the circle. Here is a regular inscribed hexagon:

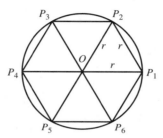

A polygon is circumscribed about a circle if every side of the polygon is tangent to the circle. Here is a regular circumscribed octogon:

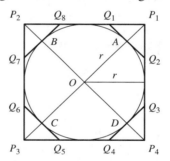

- Our study of regular polygons yielded the following approximations of the circumference of a circle of radius r ($2\pi r$) and the area of a circle (πr^2). C is the circumference and A is the area:

$$12\sqrt{2 - \sqrt{3}} \cdot r < C < 16(\sqrt{\sqrt{2} - 1}) \text{ or } 6.21r < C < 6.63r$$

$$3r^2 < A < 8(\sqrt{2} - 1)r^2 \text{ or } 3r^2 < A < 3.32r^2.$$

CHAPTER 7

Triangles and Circles

7.1 CIRCUMCIRCLES

From here until the end of the book our plane geometry chapters will be concerned with properties of triangles. Even our non-Euclidean model chapters will be concerned with analogues of Euclidean triangle theorems. In some sense the triangle is one of the simplest geometric figures imaginable. Therefore, it is remarkable that there is such a large and deep theory devoted to it. Although our treatment will not be complete, we hope to give you an appreciation of some deeper portions of this theory.

Since the rest of the book will be devoted to triangles we will try to use consistent notation throughout. As a start we will refer to the lengths of the sides of the triangle $\triangle ABC$ as $BC = a$, $AC = b$, and $AB = c$.

Our main goal of this section is to show that for any triangle $\triangle ABC$ there exists a unique circle that contains the three vertices, A, B, and C. We say that this circle

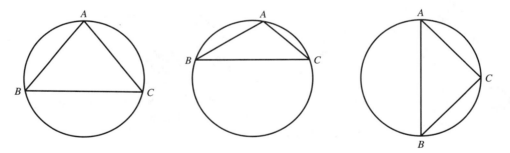

Figure 7.1: Triangles and circumcircles

is *circumscribed* about $\triangle ABC$ and call it the *circumcircle* of $\triangle ABC$. (We also say that $\triangle ABC$ is *inscribed* in the circle.) The center of the circumcircle is called the *circumcenter*, and it will be denoted O. The radius of the circumcircle is called the *circumradius*, and it will be denoted R. The key step in the proof is the following lemma.

LEMMA. Given a line segment $\overline{AB}$ and a point P, the point P will be on the perpendicular bisector of $\overline{AB}$ if and only if P is equidistant from A and B (i.e., $\overline{PA} \cong \overline{PB}$).

Proof. Let M be the midpoint of $\overline{AB}$. We will show that $\overline{PM}$ is perpendicular to $\overline{AB}$ if and only if $\overline{PA} \cong \overline{PB}$.

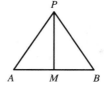

Figure 7.2: Proof of lemma

Since we are proving an if and only if statement, we have two statements to prove. First, we assume that $\overline{PA} \cong \overline{PB}$. In this case $\triangle PAM \cong \triangle PBM$ by SSS. So $\angle AMP \cong \angle BMP$, and since the sum $\angle AMP + \angle BMP$ is $180°$, each must be $90°$. Next, assume that $\overline{PM}$ is perpendicular to $\overline{AB}$. In this case since $\angle AMP \cong \angle BMP$ we use SAS to conclude that $\triangle PAM \cong \triangle PBM$. So $\overline{PA} \cong \overline{PB}$. $\qquad\square$

We now prove our main theorem.

THEOREM. Given any triangle $\triangle ABC$, there exists a unique circle circumscribed about it.

Proof. Let ℓ_1 be the perpendicular bisector of $\overline{BC}$ and let ℓ_2 be the perpendicular bisector of $\overline{AC}$. Let O be the intersection point of ℓ_1 and ℓ_2. Apply the lemma to see that since O is on ℓ_1,

$$OB = OC,$$

and, since O is on ℓ_2,

$$OA = OC.$$

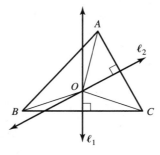

Figure 7.3: Construction of O

Combining these two equations, we see that $OA = OB = OC$. Hence, O is equidistant from the three points A, B, and C and there is a circle with center O containing these three points.

Now suppose that O' is the center of a circle that passes through each of A, B, and C. Because $O'A = O'B = O'C$ and another application of the lemma, O' must lie on both ℓ_1 and ℓ_2, and hence $O' = O$. Therefore, only one circle contains A, B, and C and the circumscribed circle we have constructed is unique.　□

As a corollary of the *proof* we get the next result.

COROLLARY. In any triangle $\triangle ABC$ the three perpendicular bisectors of the sides $\overline{AB}$, $\overline{BC}$, and $\overline{AC}$ meet in a point.

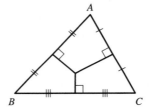

Figure 7.4: Perpendicular bisectors of sides of $\triangle ABC$

We remark that this proof is constructive. Given $\triangle ABC$ we not only know that a circumcircle exists but we also have a technique to find O and we may draw the circle.

Although O is the center of $\triangle ABC$ in the sense that it is the unique point equidistant from A, B, and C, we point out that O does not have to lie inside the triangle. You may wish to draw a few pictures or consult Fig. 7.1. O lies inside of $\triangle ABC$ if and only if $\triangle ABC$ is acute.

7.2　A THEOREM OF BRAHMAGUPTA

When you first learned about areas you were told that the the area of a rectangle was length times width. It might have occurred to you that the area of a triangle should be the product of the three sides, abc. Of course it isn't, but in this section we will prove that the product abc is related to area. We start off with a consideration of the product of two sides of a triangle.

THEOREM. Given $\triangle ABC$, let $AC = b$, $AB = c$, let R be the circumradius, and let $h = $ the length of the altitude $\overline{AD}$ from A to $\overline{BC}$. Then

$$bc = 2Rh.$$

Proof. Let A' be the point such that $\overline{AA'}$ is a diameter of the circumcircle and consider the triangles $\triangle ADB$ and $\triangle ACA'$ in Fig. 7.5. The angles $\angle A'$ and

$\angle B$ each subtend the arc $\overset{\frown}{AC}$ and are congruent, $\angle A' \cong \angle B$. Also, $\angle ACA'$ is inscribed in a semicircle, so it is a right angle. Therefore, $\angle ACA' \cong \angle ADB$. By AA, $\triangle ADB$ is similar to $\triangle ACA'$. So

$$\frac{AD}{AB} = \frac{AC}{AA'},$$

or, by cross multiplication, $AA' \cdot AD = AB \cdot AC$. But $AA' = 2R$, $AD = h$, $AB = c$, $AC = b$, and the theorem follows.   □

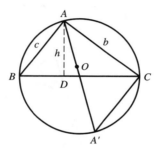

Figure 7.5: Proof of Brahmagupta's theorem

As an immediate consequence we get the following elegant theorem due to Brahmagupta. We use the letter K to denote the area of triangle $\triangle ABC$.

THEOREM. $abc = 4RK$.

Proof. By the previous theorem, $bc = 2Rh$. Now multiply both sides by a to get

$$abc = 2Rah.$$

But $K = \frac{1}{2}ah$ and the theorem follows.   □

7.3 INSCRIBED CIRCLES

This section is analogous to the first section of this chapter. We will prove that given any triangle $\triangle ABC$ there is a unique circle inscribed in $\triangle ABC$ called the *incircle* of $\triangle ABC$. (A circle is said to be inscribed in a triangle if each of the three sides is tangent to the circle. The triangle is said to be circumscribed about the circle.) The center of the incircle (inscribed circle) is called the incenter and is denoted I. The radius is called the inradius and is denoted r. In order to prove the existence and uniqueness of incircles, we need a lemma on angle bisectors.

LEMMA. Let $\angle A$ be any angle and let P be a point in the interior of $\angle A$. Then P is on the bisector of $\angle A$ if and only if P is equidistant from the sides of $\angle A$.

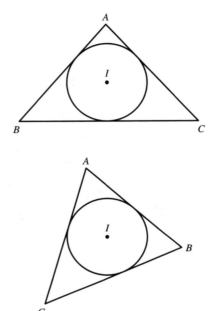

Figure 7.6: Triangles and Incircles

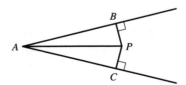

Figure 7.7: Proof of lemma

Proof. Let $\overline{PB}$ and $\overline{PC}$ be the perpendiculars from P to the sides of $\angle A$. We need to show that $\overline{PB} \cong \overline{PC}$ if and only if $\angle PAB \cong \angle PAC$.

First, assume that $\overline{PB} \cong \overline{PC}$. Then $\triangle PAB \cong \triangle PAC$ by SSA for right triangles. Since $\angle PAB$ and $\angle PAC$ are corresponding parts, they must be congruent. Next, assume that $\angle PAB \cong \angle PAC$. Then by SAA $\triangle PAB$ and $\triangle PAC$ are congruent for they have two angles congruent and share a side. Hence $\overline{PB} \cong \overline{PC}$. $\square$

THEOREM. Given any triangle $\triangle ABC$, there is a unique circle inscribed in it.

Proof. Let ℓ_1 be the bisector of $\angle A$ and let ℓ_2 be the bisector of $\angle B$. Let ℓ_1 and ℓ_2 intersect at the point I and apply the lemma: Since I is on the bisector of $\angle A$, the distance from I to $\overline{AB}$ equals the distance from I to $\overline{AC}$. Similarly, since I is on the bisector of $\angle B$, the distance from I to $\overline{BC}$ equals the distance from I to $\overline{AB}$. Combining these, we see that I is equidistant from the three sides $\overline{AB}$, $\overline{BC}$, and $\overline{AC}$. So there is a circle with center I that will be tangent to all three. To show that this circle is the unique incircle, suppose that P is the center of a circle D inscribed in $\triangle ABC$. Drop segments from P to E and F, the points of tangency of the lines $\overleftrightarrow{AB}$ and $\overleftrightarrow{AC}$, respectively, with D. We know from previous work that $\overline{PE} \perp \overline{AB}$ and $\overline{PF} \perp \overline{AC}$; moreover, $PE = PF$ since both are radii of D. Applying our lemma, we see that P is on ℓ_1, the bisector of $\angle A$. Similarly, P is also on ℓ_2, which implies that $P = I$ and that the first circle we constructed is unique. $\square$

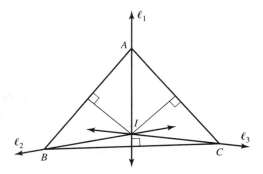

Figure 7.8: Construction of I

As a corollary of the proof we get the next result.

COROLLARY. Given any triangle $\triangle ABC$, the bisectors of the three angles meet in one point.

We point out that our proof here is also constructive. Given $\triangle ABC$, we can construct I by finding the meeting point of the angle bisectors and we can use I to construct the incircle by dropping perpendiculars. One way in which I is better behaved

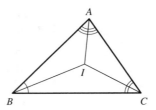

Figure 7.9: Bisectors of angles of $\triangle ABC$

than O is that I always lies inside the triangle. This fact has an interesting implication for area.

THEOREM. If $K =$ the area of $\triangle ABC$, $r =$ the inradius, and $s = \frac{1}{2}$ times the perimeter, then $K = sr$.

Proof. Draw IA, IB, and IC as in Fig. 7.9. It is clear that the area of $\triangle ABC$ is the sum of the areas of $\triangle ABI$, $\triangle BCI$, and $\triangle ACI$. Consider $\triangle ABI$. The base is $\overline{AB}$, which has length c, and the height is the distance from I to AB, which is r. So area $\triangle ABI = \frac{1}{2}cr$. Likewise, $\triangle BCI$ has area $\frac{1}{2}ar$ and $\triangle ACI$ has area $\frac{1}{2}br$. By addition,

$$K = \frac{1}{2}ar + \frac{1}{2}br + \frac{1}{2}cr$$

$$= \frac{1}{2}(a + b + c)r$$

$$= sr.$$

□

7.4   AN OLD CHESTNUT (THE STEINER-LEHMUS THEOREM)

In this section we will prove the following theorem.

THEOREM. In $\triangle ABC$ let the bisector of $\angle B$ meet $\overline{AC}$ at D and let the bisector of $\angle C$ meet $\overline{AB}$ at E. If $\overline{BD} \cong \overline{CE}$, then $\angle B \cong \angle C$.

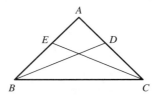

Figure 7.10: Triangle with equal angle bisectors

In addition to the intrinsic interest of this theorem, it is also famous as a puzzle. The statement of the theorem is fairly easy and the proof is not that difficult to follow, but most people find it quite hard to prove. You may want to try it yourself before you read this section. If you become a teacher you may want to treat your students to this problem. (One of the authors spent a number of his younger years wondering about this problem after a teacher was kind enough to suggest it.) We point out that if instead of taking $\overline{BD}$ and $\overline{CE}$ to be angle bisectors we take them to be altitudes or medians, the corresponding theorems are also true and are not as hard to prove. Here is the proof.

Proof. Our proof will be by contradiction. We will assume that $\overline{BD} \cong \overline{CE}$ but that $\angle B$ is larger than $\angle C$.

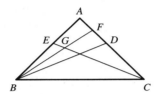

Figure 7.11: Proof that $\angle B \cong \angle C$

Since $\angle B$ is greater than $\angle C$, $\angle ABD$, which is half of $\angle B$, is greater than half of $\angle C$. So we can find a point F strictly between A and D such that $\angle DBF = \frac{1}{2}\angle C$. Let G be the intersection point of $\overline{CE}$ and $\overline{BF}$.

Now consider $\triangle BDF$ and $\triangle CGF$. Each has $\angle BFC$ for one angle. Also, $\angle FCG \cong \angle FBD$, since each is $\frac{1}{2}\angle C$. Hence, by AA, $\triangle BDF \sim \triangle CGF$ and thus

$$\frac{CG}{BD} = \frac{CF}{BF}.$$

By hypothesis $CE = BD$, which forces $CG < BD$. This implies that $1 > \frac{CG}{BD}$ and, by the preceding equation, that

$$BF > CF.$$

Now that we have shown that $BF > CF$ we can get a contradiction, because we have another line of reasoning that will show the opposite. Consider $\triangle BFC$. The angle $\angle FBC = \frac{1}{2}\angle B + \frac{1}{2}\angle C > \angle C$, because we assumed that $\angle B > \angle C$. But, in any triangle, a larger angle must be opposite a larger side. So $CF > BF$. This is a contradiction and so our initial assumption is impossible. Assuming $\angle C > \angle B$ will yield the same contradiction. Thus the theorem is proved.   □

7.5 ESCRIBED CIRCLES

Before discussing escribed circles, it is worthwhile to define external bisectors of angles.

Let $\angle BAC$ be any angle with bisector $\overleftrightarrow{AD}$. Extend one of the sides of the angle, say $\overrightarrow{AB}$, in the other direction to form $\angle B'AC$, the supplementary angle. Then the line $\overleftrightarrow{AE}$ which bisects this angle is called the *external bisector* of $\angle BAC$. External bisectors have two simple properties that will be important to us.

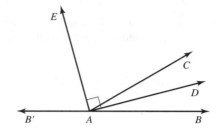

Figure 7.12: External bisector of $\angle BAC$

LEMMA. Let $\angle BAC$ have internal bisector $\overleftrightarrow{AD}$ and external bisector $\overleftrightarrow{AE}$. Then
(1) $\overleftrightarrow{AD}$ is perpendicular to $\overleftrightarrow{AE}$.
(2) The points of $\overleftrightarrow{AE}$ are equidistant from $\overleftrightarrow{AB'}$ and $\overleftrightarrow{AC}$.

Proof. (1)

$$\angle DAE = \angle DAC + \angle CAE$$
$$= \frac{1}{2}\angle BAC + \frac{1}{2}\angle CAB'$$
$$= \frac{1}{2}\angle BAC + \frac{1}{2}(180° - \angle BAC)$$
$$= 90°.$$

(2) The points of $\overleftrightarrow{AE}$ are on the bisector of $\angle B'AC$ and so they are equidistant from $\overleftrightarrow{AC}$ and $\overleftrightarrow{AB'} = \overleftrightarrow{AB}$. □

We are now in a position to generalize all of the results of Section 7.3. In $\triangle ABC$ draw the external bisectors of $\angle B$ and $\angle C$ and label the point of intersection I_a. By the lemma, since I_a is on the external bisector of $\angle B$, the distance from I_a to $\overleftrightarrow{BC}$ equals the distance from I_a to $\overleftrightarrow{AB}$ and since, I_a is on the external bisector of $\angle C$, the distance from I_a to $\overleftrightarrow{BC}$ equals the distance from I_a to $\overleftrightarrow{AC}$. Hence I_a (like I) is equidistant from the three lines $\overleftrightarrow{AB}$, $\overleftrightarrow{BC}$, and $\overleftrightarrow{AC}$. There are two conclusions we may draw from this observation, and we leave the details of proof to you.

THEOREM. Let $\triangle ABC$ be any triangle. Then the external bisector of $\angle B$, the external bisector of $\angle C$, and the internal bisector of $\angle A$ all meet in a point I_a.

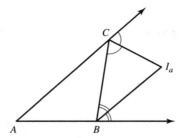

Figure 7.13: Meeting point of two external bisectors

Moreover, there is a circle with center I_a tangent to the three lines $\overleftrightarrow{AB}$, $\overleftrightarrow{BC}$, and $\overleftrightarrow{AC}$.

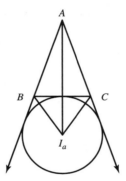

Figure 7.14: The excenter I_a and the angle bisectors

The circle we constructed in this manner is said to be an *escribed circle* for $\triangle ABC$, the point I_a is called an *excenter* , and the radius of the circle, which we denote r_a, is called an *exradius*. The triangle $\triangle ABC$ has two more excircles, one touching $\overline{AC}$ and one touching $\overline{AB}$. We denote their centers by I_b and I_c and their radii as r_b and r_c, respectively. Our next theorem generalizes the area formula of Section 7.3 to the case of excircles.

THEOREM. Let $K =$ the area of $\triangle ABC$, $r_a =$ the radius of the escribed circle opposite $\angle A$, $BC = a$, and $s =$ the semiperimeter. Then $K = (s - a)r_a$.

Proof. The area of $\triangle ABC =$ the area of $\triangle ABI_a +$ the area of $\triangle ACI_a -$ the area of $\triangle BCI_a$ (See Fig. 7.15). Taking $\overline{AC}$ as the base, we see that $\triangle ACI_a$ has area $\frac{1}{2}br_a$ and $\triangle BCI_a$ has area $\frac{1}{2}ar_a$. Hence the area of $\triangle ABC = \frac{1}{2}cr_a + \frac{1}{2}br_a - \frac{1}{2}ar_a = \frac{1}{2}r_a(b + c - a) = \frac{1}{2}r_a(a + b + c - 2a) =$

$$\tfrac{1}{2}r_a \cdot (2s - 2a) = \tfrac{1}{2}r_a \cdot 2 \cdot (s - a) = r_a(s - a). \qquad \square$$

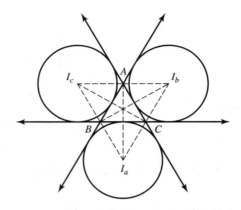

Figure 7.15: The three excenters

This theorem, together with the other area theorems, has some striking corollaries.

COROLLARY. If r is the inradius of $\triangle ABC$ and K is its area, then $rr_a r_b r_c = K^2$.

Proof. By the theorem, $r_a = \frac{K}{s-a}$. Likewise $r_b = \frac{K}{s-b}$ and $r_c = \frac{K}{s-c}$. Also, since $K = rs$, $r = \frac{K}{s}$. Hence

$$rr_a r_b r_c = \frac{K^4}{s(s-a)(s-b)(s-c)} = \frac{K^4}{K^2} = K^2$$

by Heron's formula. □

COROLLARY. $\frac{1}{r} = \frac{1}{r_a} + \frac{1}{r_b} + \frac{1}{r_c}$

Proof.

$$\frac{1}{r_a} + \frac{1}{r_b} + \frac{1}{r_c} = \frac{s-a}{K} + \frac{s-b}{K} + \frac{s-c}{K}$$

$$= \frac{3s - (a+b+c)}{K}$$

$$= \frac{3s - 2s}{K} = \frac{s}{K} = \frac{1}{r}$$

□

7.6 EULER'S THEOREM

In this section we will prove a celebrated theorem due to Euler that calculates the distance from the incenter I to the circumcenter O. The proof is long but beautiful.

THEOREM. $OI^2 = R(R - 2r)$.

The proof will depend upon a two part lemma. In the circumcircle of $\triangle ABC$ pictured in Fig. 7.16 we will denote by M the midpoint of arc $\overset{\frown}{BC}$.

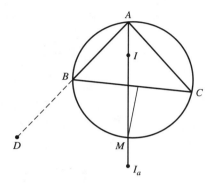

Figure 7.16: Midpoint of $\overset{\frown}{BC}$

LEMMA. (1) M lies on the perpendicular bisector of $\overline{BC}$ and the bisector of $\angle A$.
(2) There is a circle with center M that contains the points I, B, C, and I_a.

Proof. (1) Since $\overset{\frown}{CM} \cong \overset{\frown}{MB}$, $\overline{CM} \cong \overline{MB}$. But since M is equidistant from B and C, it must lie on the perpendicular bisector of $\overline{BC}$. Also, $\angle BAM \cong \angle CAM$ since they subtend equal arcs.

(2) First notice that $\overline{IB}$ bisects $\angle B$ and that $\overline{BI_a}$ bisects $\angle CBD$, where D is a point on $\overline{AB}$ extended. Since $\angle ABD$ is a straight angle, the angle $\angle IBI_a$ is a right angle. Similarly, $\angle ICI_a$ is a right angle. Hence, if we draw a circle with $\overline{II_a}$ as diameter it will contain the points B and C. M must be the center of this circle because it lies on the intersection of a diameter with the perpendicular bisector of a chord. $\qquad\square$

Proof of Euler's Theorem. Extend $\overline{OI}$ to a chord $\overline{PQ}$ of the circumcircle. The line $\overline{AI}$ extends to the chord $\overline{AM}$. Hence

$$PI \cdot IQ = AI \cdot IM. \tag{7.1}$$

The rest of the proof will consist of an examination of each side of equation 7.1.

We first consider the left-hand side. $PI = OP - OI = R - OI$ and $IQ = IO + OQ = (R + OI)$. Hence $\text{PI} \cdot IQ = (R - OI)(R + OI) = R^2 - OI^2$.

To evaluate the right-hand side we draw $\overline{IZ}$ perpendicular to $\overline{AB}$ and extend $\overline{MO}$ to a diameter $\overline{MM'}$. We claim that $\triangle AIZ \sim \triangle M'MC$. This is because $\angle Z = \angle C = 90°$ and $\angle ZAI = \angle CM'M = \frac{1}{2}\angle A$. Hence $\frac{AI}{IZ} = \frac{M'M}{CM}$. What can

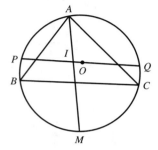

Figure 7.17: Proof of Euler's theorem

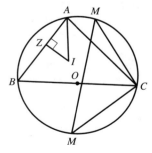

Figure 7.18: Proof of Euler's theorem (cont.)

we say about these lengths? By the lemma $CM = IM$, since they are radii of the same circle. $\overline{MM'}$ is a diameter of the circumcircle and has length $2R$. IZ is the distance from the incenter to a side of the triangle, so it is r. Making these three substitutions we see that $\frac{AI}{r} = \frac{2R}{IM}$, so $AI \cdot IM = 2Rr$.

Now we substitute the results of each of the preceeding paragraphs into equation 7.1 to get $R^2 - OI^2 = 2Rr$ or $OI^2 = R^2 - 2Rr = R(R - 2r)$.   □

COROLLARY. $R \geq 2r$

PROBLEMS

7.1. Assume that $\triangle ABC$ and $\triangle DEF$ are similar with ratio k. Prove that each of the circumradius, the inradius, and the exradii of $\triangle DEF$ are k times the corresponding parts of $\triangle ABC$.

7.2. In each part determine the missing information about $\triangle ABC$.
   (a) $a = 4, b = 4, c = 2, K = ?, R = ?, r = ?, r_a = ?, r_b = ?, r_c = ?$
   (b) $r_a = 2, r_b = 3, r_c = 6, r = ?, K = ?, R = ?, a = ?, b = ?, c = ?$

7.3. Calculate the angle between the altitude $\overline{AD}$ and the circumradius $\overline{AO}$ in terms of the base angles, $\angle B$ and $\angle C$. (See Fig. 7.5.)

7.4. Let $\triangle ABC$ be an equilateral triangle and let P be a point inside. Prove that the sum of the distances from P to the three sides of $\triangle ABC$ is equal to the altitude $\overline{AD}$. [Hint: The area of $\triangle ABC$ is the sum of the areas of $\triangle ABP$, $\triangle BCP$, and $\triangle ACP$.]

*7.5. Prove that a quadrilateral $ABCD$ can be inscribed in a circle if and only if $\angle A + \angle C = 180°$.

7.6. Prove that if a quadrilateral $ABCD$ is circumscribed about a circle, then the area of $ABCD$ is one-half times the radius of the circle times the perimeter.

7.7. Prove that every rhombus has an inscribed circle.

7.8. Given $\triangle ABC$ such that the bisector of $\angle B$ meets $\overline{AC}$ at D and the bisector of $\angle C$ meets $\overline{AB}$ at E, prove that if $\angle B > \angle C$, then $BD < CE$.

*7.9. **(a)** Assume that the incircle of $\triangle ABC$ is tangent to $\overline{BC}$ at X, $\overline{AC}$ at Y, and $\overline{AB}$ at Z. Find the lengths of $\overline{BX}, \overline{XC}, \overline{AY}, \overline{YC}, \overline{AZ}$, and $\overline{BZ}$ in terms of a, b, and c

   (b) Assume that the excircle with center I_a is tangent to $\overline{BC}$ at X_a, $\overrightarrow{AC}$ at Y_a, and $\overrightarrow{AB}$ at Z_a. Find the lengths of $\overline{AZ_a}, \overline{AY_a}, \overline{BX_a}, \overline{CY_a}, \overline{BX_a}$, and $\overline{CX_a}$.

7.10. Let the lengths of the three altitudes be h_a, h_b, and h_c. Prove

$$\frac{1}{h_a} + \frac{1}{h_b} + \frac{1}{h_c} = \frac{1}{r}.$$

7.11. Prove that the circle with diameter $\overline{I_b I_c}$ has center on the circumcircle and contains the points B and C.

*7.12. Prove that $R = 2r$ if and only if $\triangle ABC$ is equilateral.

7.13. Prove: $OI_a^2 = R(R + 2r_a)$.

7.14. (a) Given r_a, r_b, r_c, construct r.

   (b) Given r, r_a, r_b, r_c, construct a square with area K.

   (c) Given I_a, I_b, I_c, I (or any three of them), construct $\triangle ABC$.

CHAPTER SUMMARY

- Any triangle has a unique circumcircle with circumcenter O.

- The circumcircle of a triangle is constructed by drawing the perpendicular bisectors of the sides of the triangle, which always meet in a point. This point is
  the center of the circumcircle, and the distances from it to the three vertices of
  the triangle will be lengths of the radii.

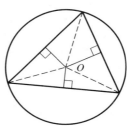

- The figure below yields the following relationship: $bc = 2Rh$, where R is the
  circumradius; also, $abc = 4RK$ follows from algebra, as the theorem of Brahmagupta.

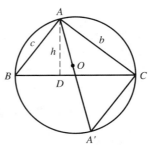

- Any triangle has a unique inscribed circle with incenter I.

- The incircle is constructed by drawing the angle bisectors of the triangle in question, which always meet in a point. That point is the center of the incircle, and
  the segments from it perpendicular to each of the three sides of the triangle are
  the radii.

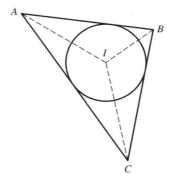

- In the following figure, $\overline{CE}$ and $\overline{BD}$ are the angle bisectors of $\angle C$ and $\angle B$ respectively. If $\overline{BD} \cong \overline{CE}$, then $\angle B \cong \angle C$.

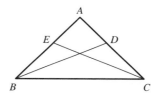

- The external bisector of $\angle B$, the external bisector of $\angle C$, and the internal bisector of $\angle A$ meet in a point I_a. There is a circle, called an escribed circle with center I_a tangent to $\overleftrightarrow{AB}$, $\overleftrightarrow{BC}$, and $\overleftrightarrow{AC}$ as shown. Parallel arguments hold for I_b and I_c.

- Moreover,

$$K = (s - a)r_a,$$

$$K = (s - b)r_b,$$

$$K = (s - c)r_c,$$

$$K = sr,$$

$$rr_a r_b r_c = K^2,$$

$$\frac{1}{r_a} + \frac{1}{r_b} + \frac{1}{r_c} = \frac{1}{r}.$$

The preceeding are formulas derived in this chapter where s is the semiperimeter, K is the area of $\triangle ABC$, r is the inradius, r_a is the radius of the escribed circle opposite $\angle A$, r_b is the radius of the escribed circle opposite $\angle B$, and r_c is the radius of the escribed circle opposite $\angle C$.

- In the figure, M lies on the perpendicular bisector of $\overline{BC}$, and the circle with center M would contain points I, B, C, and I_a.

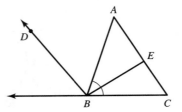

- Also, $OI^2 = R(R - 2r)$, where R is the circumradius, r is the inradius, and OI is the distance from the incenter I to the circumcenter O.

CHAPTER 8

Medians

8.1 CENTER OF GRAVITY

In $\triangle ABC$ let the midpoints of sides $\overline{BC}$, $\overline{AC}$, and $\overline{AB}$ be A', B', and C', respectively. The line segments $\overline{AA'}$, $\overline{BB'}$, and $\overline{CC'}$ connecting the three vertices to the midpoints of the opposite sides are called the *medians* of $\triangle ABC$.

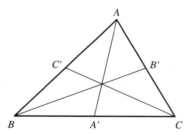

Figure 8.1: Medians and centroid of $\triangle ABC$

Our main theorem in this section states that the three medians meet in a point G called the *centroid* or *center of gravity* of $\triangle ABC$. The proof will be based upon a careful study of the intersection of two medians.

LEMMA. Let B' be the midpoint of $\overline{AC}$ and C' be the midpoint of $\overline{AB}$. Then $\overline{B'C'}$ is parallel to $\overline{BC}$ and is half as long.

Proof. The triangles $\triangle ABC$ and $\triangle AC'B'$ are similar by the SAS theorem for similar triangles because $\angle BAC \cong \angle C'AB'$, $AC' = \frac{1}{2}AB$, and $AB' = \frac{1}{2}AC$.

Hence $\angle AC'B' \cong \angle ABC$ and $\overline{BC} \| \overline{B'C'}$, since they make equal corresponding angles with the transversal $\overleftrightarrow{AC'}$. Finally, $\frac{B'C'}{BC} = \frac{AC'}{AB} = \frac{1}{2}$. $\qquad\square$

Let us label the intersection point of $\overline{BB'}$ and $\overline{CC'}$ as G. We don't yet know that G is also on $\overline{AA'}$ We first prove, as an intermediate step, that G is the trisection point of $\overline{BB'}$.

LEMMA. $GB' = \frac{1}{3}BB'$

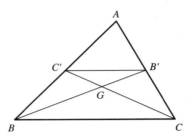

Figure 8.2: Two medians in a triangle

Proof. Since $\overline{B'C'}$ is parallel to $\overline{BC}$, $\angle B'BC \cong \angle BB'C'$ and $\angle CC'B' \cong \angle C'CB$ (Fig. 8.2) since they are alternate interior angles. Hence, by AA, $\triangle GBC \sim \angle GB'C'$. So

$$\frac{GB}{GB'} = \frac{BC}{B'C'} = 2.$$

Hence $GB = 2GB'$. If we add GB' to both sides we see that $BB' = 3GB'$. $\square$

Using this lemma, the proof that the three medians meet in a point is short, although if you are not used to this type of argument it is worth reading carefully.

THEOREM. In $\triangle ABC$ the three medians $\overline{AA'}$, $\overline{BB'}$, and $\overline{CC'}$ meet in a point.

Proof. Let G be the point of intersection of $\overline{BB'}$ and $\overline{CC'}$. We need to show that G also lies on $\overline{AA'}$. In order to do this we let $G_1 =$ the intersection point of $\overline{AA'}$

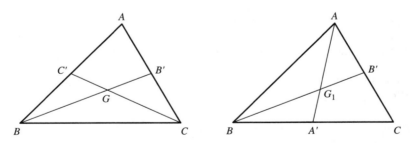

Figure 8.3: Proof that medians meet at G

and $\overline{BB'}$ and we will show that $G = G_1$. By the lemma $B'G = \frac{1}{3}BB'$. However, we can also apply the lemma to G_1. G_1 is the intersection point of two medians in a triangle and so it must also trisect them. Hence, $B'G_1 = \frac{1}{3}BB'$ But $\overline{BB'}$ only has one trisection point closer to B' so $G = G_1$. $\square$

As a corollary to the proof we have the following fact.

COROLLARY. The center of gravity trisects each of the medians.

8.2 LENGTH FORMULAS

In this section we will calculate the lengths of the medians of $\triangle ABC$. We will denote these lengths by $AA' = m_a$, $BB' = m_b$, and $CC' = m_c$.

THEOREM. $2m_a^2 = b^2 + c^2 - \frac{1}{2}a^2$.

Proof. Let $\overline{AD}$ be the altitude. We assume for convenience that D is between B and A', as shown in Fig. 8.4. Now apply the geometric law of cosines to the

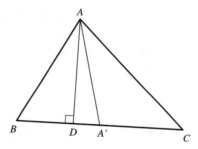

Figure 8.4: Length of $\overline{AA}'$

triangles, $\triangle ABA'$ and $\triangle ACA'$ to obtain

$$AA'^2 = AB^2 + BA'^2 - 2BD \cdot BA'$$

and

$$AA'^2 = AC^2 + CA'^2 - 2CD \cdot CA',$$

or

$$m_a^2 = c^2 + (\frac{1}{2}a)^2 - 2(BD)(\frac{1}{2}a)$$

and

$$m_a^2 = b^2 + (\frac{1}{2}a)^2 - 2(CD)(\frac{1}{2}a).$$

Adding, we get

$$2m_a^2 = b^2 + c^2 + \frac{1}{2}a^2 - a(CD + BD)$$

$$= b^2 + c^2 + \frac{1}{2}a^2 - a^2 = b^2 + c^2 - \frac{1}{2}a^2,$$

as claimed.   □

Of course, there are similar formulas for the other two medians:

$$2m_b^2 = a^2 + c^2 - \frac{1}{2}b^2$$

and

$$2m_c^2 = a^2 + b^2 - \frac{1}{2}c^2.$$

With a bit of algebra we can deduce a number of striking formulas from these.

COROLLARY. (a) $m_a^2 + m_b^2 + m_c^2 = \frac{3}{4}(a^2 + b^2 + c^2)$

(b) $GA^2 + GB^2 + GC^2 = \frac{1}{3}(a^2 + b^2 + c^2)$

(c) $GA'^2 + GB'^2 + GC'^2 = \frac{1}{12}(a^2 + b^2 + c^2)$

8.3   COMPLEMENTARY AND ANTICOMPLEMENTARY TRIANGLES

Let $\triangle ABC$ be a triangle and let the midpoints of the sides be A', B', and C'. Then the triangle with these three points for vertices is called the *complementary* or *medial* triangle of $\triangle ABC$. Based upon our work in Section 8.1, we already know a lot about $\triangle A'B'C'$.

THEOREM. Given $\triangle ABC$ with complementary triangle $\triangle A'B'C'$,

(1) The sides of $\triangle A'B'C'$ are parallel to the sides of $\triangle ABC$.

(2) $\triangle ABC$ and $\triangle A'B'C'$ are similar with ratio $\frac{1}{2}$.

(3) $\triangle ABC$ and $\triangle A'B'C'$ have the same centroid G.

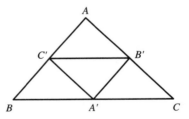

Figure 8.5: Complementary triangle of $\triangle ABC$

Proof. (1) We showed in Section 8.1 that $\overline{BC}$ and $\overline{B'C'}$ are parallel. The same proof shows that $\overline{AB}$ and $\overline{A'B'}$ are parallel and $\overline{AC}$ and $\overline{A'C'}$ are parallel.

(2) We also proved in Section 8.1 that $B'C' = \frac{1}{2} \cdot BC$. The same proof shows that $A'B' = \frac{1}{2} \cdot AB$ and $A'C' = \frac{1}{2}AC$. Hence, $\triangle ABC \sim \triangle A'B'C'$ by SSS for similar triangles.

(3) Let $\overline{AA'}$ intersect $\overline{B'C'}$ at the point M. Consider the triangles $\triangle AA'B$ and $\triangle AMC'$. Since $\overline{B'C'} \| \overline{BC}$, $\angle C' \cong \angle B$ by the corresponding angles theorem. Likewise, $\angle AMC' \cong \angle AA'B$. So the two triangles must be similar by AA. Hence

$$\frac{C'M}{BA'} = \frac{AC'}{AB} = \frac{1}{2}$$

which implies that $C'M = \frac{1}{2}BA'$. But $BA' = \frac{1}{2}BC$ so $BA' = B'C'$. Putting these together we see that $C'M = \frac{1}{2}B'C'$ and that M is the midpoint of $\overline{B'C'}$. With this observation in hand the rest of the proof follows easily. $\overline{AA'}$ is a median of $\triangle ABC$, $\overline{A'M}$ is a median of $\triangle A'B'C'$, and $\overleftrightarrow{AA'} = \overleftrightarrow{A'M}$. Similar statements are true for the other two medians. Hence, the point of intersection of the three medians of $\triangle ABC$ is the same as the point of intersection of the three medians of $\triangle A'B'C'$. $\qquad\square$

Next, given a triangle $\triangle ABC$, we define the *anticomplementary* triangle of $\triangle ABC$ to be the unique triangle $\triangle A''B''C''$ with the properties that $\overline{B''C''}$ contains the point A and is parallel to $\overline{BC}$, $\overline{A''C''}$ contains the point B and is parallel to $\overline{AC}$, and $\overline{A''B''}$ contains the point C and is parallel to $\overline{AB}$.

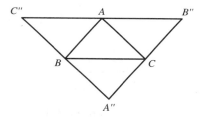

Figure 8.6: Anticomplementary triangle of $\triangle ABC$

THEOREM. If a triangle $\triangle ABC$ has anticomplementary triangle $\triangle A''B''C''$, then $\triangle ABC$ is the complementary triangle of $\triangle A''B''C''$.

Proof. To prove this theorem it is worthwhile to recall the definition of a complementary triangle. In order to prove that $\triangle ABC$ is the complementary triangle of $\triangle A''B''C''$ we need to show (see Fig. 8.6) that A is the midpoint of $\overline{B''C''}$, B is the midpoint of $\overline{A''C''}$ and C is the midpoint of $\overline{A''B''}$. We will only prove that A is the midpoint of $\overline{B''C''}$ since the other two cases are similar.

Because $\overline{AB''} \| \overline{BC}$ and $\overline{AB} \| \overline{B''C}$, the quadrilateral $BCB''A$ must be a parallelogram. We know that opposite sides of a parallelogram are congruent, so

$$\overline{AB''} \cong \overline{BC}.$$

Likewise, since $\overline{AC''} \| \overline{BC}$ and $\overline{BC''} \| \overline{AC}$, the quadrilateral $BCAC''$ is also a parallelogram and consequently

$$\overline{AC''} \cong \overline{BC}.$$

Comparing these two congruences, we see that $\overline{C''A} \cong \overline{AB''}$, which means that A must be the midpoint of $\overline{B''C''}$. This is what we wanted to show. $\qquad\square$

COROLLARY.   **1.** $\triangle ABC$ and $\triangle A''B''C''$ are similar with ratio 2.

2. $\triangle ABC$ and $\triangle A''B''C''$ have the same centroid.

PROBLEMS

8.1. If $\overline{AA'}$ is a median of $\triangle ABC$, prove that it is not (the extension of) a median in any other triangle $\triangle ADE$ with D on $\overrightarrow{AB}$ and E on $\overrightarrow{AC}$.

8.2. In triangle $\triangle ABC$ let C_1 on $\overline{AB}$ be such that $AC_1 = \frac{1}{3}AB$ and let B_1 on $\overline{AC}$ be such that $AB_1 = \frac{1}{3}AC$. Prove:

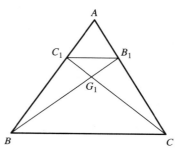

Figure 8.7: Exercise 2

(a) $\overline{B_1C_1} \| \overline{BC}$.

(b) $B_1C_1 = \frac{1}{3}BC$

(c) If $\overline{BB_1}$ and $\overline{CC_1}$ intersect at point G_1, then $B_1G_1 = \frac{1}{4}BB_1$ and $C_1G_1 = \frac{1}{4}CC_1$.

8.3. If, in $\triangle ABC$, a is the length of side $\overline{AB}$, b the length of $\overline{AC}$, and m_a and m_b the lengths of the medians from A and B, respectively, then prove that $a \geq b$ implies $m_a \leq m_b$.

8.4. Use the triangle inequality on $\triangle ABA'$ and $\triangle ACA'$ to prove that $2m_a > b + c - a$. What does this imply about $m_a + m_b + m_c$?

*8.5. Prove that $m_a < \frac{1}{2}(b + c)$. What does this imply about $m_a + m_b + m_c$?

*8.6. Given a triangle $\triangle ABC$, a *derived triangle* is a triangle whose sides are congruent to the medians of $\triangle ABC$ (i.e., a triangle $\triangle DEF$ such that $DE = m_a$, $EF = m_b$ and $DF = m_c$). Prove:

(a) $\triangle ABC$ is isosceles if and only if any derived triangle is isosceles.

(b) A derived triangle of a derived triangle of $\triangle ABC$ is similar to $\triangle ABC$ with ratio $\frac{3}{4}$.

(c) What can you say about the derived triangle of the derived triangle of $\triangle ABC$?

(d) Assume that $a \geq b \geq c$. Then the derived triangle is similar to $\triangle ABC$ if and only if $2b^2 = a^2 + c^2$.

8.7. Given points B and C and a length k, prove that the set of points in the plane $\{A | AB^2 + AC^2 = k^2\}$ is a circle whose center is the midpoint of $\overline{BC}$.

8.8. What is the anticomplementary triangle of the complementary triangle of $\triangle ABC$?

8.9. Prove that if $\triangle ABC$ has complementary triangle $\triangle A'B'C'$ with centroid G, then the six triangles $\triangle GA'C$, $\triangle GCB'$, $\triangle GB'A$, $\triangle GAC'$, $\triangle GC'B$, and $\triangle GBA'$ all have equal areas.

8.10. If $\triangle ABC$ has area 1, calculate the area of each of the following:

   (a) $\triangle ABG$

   (b) $\triangle A''B''C''$

   (c) $\triangle B'CG$

   (d) quadrilateral $AB''CB$

*8.11. Let $ABCD$ be any quadrilateral. Let the midpoints of the sides be E, F, G, and H. Prove that $EFGH$ is a parallelogram.

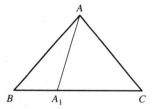

Figure 8.8: Exercise 12

8.12. In $\triangle ABC$, let $a = BC$, $b = AC$, $c = AB$. Assume that A_1 is a trisection point of $\overline{BC}$ with $A_1B = (1/3)a$. Prove: $3AA_1^2 = b^2 + 2c^2 - (2/3)a^2$.

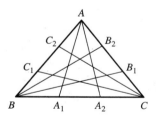

Figure 8.9: Exercise 13

8.13. Let the six trisection points of the sides of $\triangle ABC$ be $A_1, A_2, B_1, B_2, C_1, C_2$ as shown in Fig. 8.9. Generalize the previous exercise to give formulas for AA_2^2, BB_1^2, BB_2^2, CC_1^2, CC_2^2. Calculate $AA_1^2 + AA_2^2 + BB_1^2 + BB_2^2 + CC_2^2$ (proof of formulas not required).

8.14. Generalize the results of Exercises 12 and 13 to the case in which the sides of $\triangle ABC$ are divided into four equal parts.

8.15. See "Proofs without Words: The Triangle of the Medians Has Three-Fourths the Area of the Original Triangle", by Norbert Hungerbühler, in Mathematics Magazine, vol. 72(April 1999), p. 142, and write out a complete proof.

CHAPTER SUMMARY

- In the triangle A', B', and C' are the midpoints of the sides. The line segments $\overline{AA'}$, $\overline{BB'}$ and $\overline{CC'}$ are called the medians of $\triangle ABC$; they always meet at a point G, called the centroid or the center of gravity. Also, $B'C' \| \overline{BC}$, $B'C' = \frac{1}{2}BC$, and the centroid trisects the medians; for example, $B'G = \frac{1}{3}BG$.

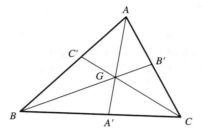

- Here are some corresponding median length formulas:

$$2m_a^2 = b^2 + c^2 - \frac{1}{2}a^2$$

$$2m_b^2 = a^2 + c^2 - \frac{1}{2}b^2$$

$$2m_c^2 = a^2 + b^2 - \frac{1}{2}c^2$$

where m_a is the length of $\overline{AA'}$, m_b is the length of $\overline{BB'}$ and m_c is the length of $\overline{CC'}$. Using algebra, we can also deduce the following.

$$m_a^2 + m_b^2 + m_c^2 = \frac{3}{4}(a^2 + b^2 + c^2)$$

$$GA^2 + GB^2 + GC^2 = \frac{1}{3}(a^2 + b^2 + c^2)$$

$$GA'^2 + GB'^2 + GC'^2 = \frac{1}{12}(a^2 + b^2 + c^2).$$

- In the figure $\triangle A'B'C'$ is called the complementary triangle of $\triangle ABC$ and has these properties:

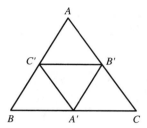

1. The three sides of $\triangle A'B'C'$ are parallel to the sides of $\triangle ABC$.
2. $\triangle ABC$ and $\triangle A'B'C'$ are similar with ratio $\frac{1}{2}$.
3. $\triangle ABC$ and $\triangle A'B'C'$ have the same centroid.

- In the figure $\triangle A''B''C''$ is called the anticomplementary triangle of $\triangle ABC$, and has the following properties:

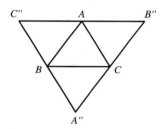

1. $\triangle ABC$ is the complementary triangle of $\triangle A''B''C''$.
2. $\triangle ABC$ and $\triangle A''B''C''$ are similar with ratio 2.
3. $\triangle ABC$ and $\triangle A''B''C''$ have the same centroid.

CHAPTER 9

Altitudes

9.1 THE ORTHOCENTER

In the previous two chapters we proved that in any triangle $\triangle ABC$ the three medians meet in a point G, the three angle bisectors meet in a point I, and the three perpendicular bisectors of the sides meet in a point O. In this section we will prove the corresponding theorem for the altitudes, namely, that the three altitudes meet in a point. This point is called the *orthocenter* of $\triangle ABC$, and we will denote it by H. We point

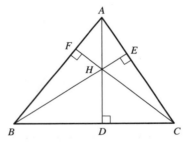

Figure 9.1: The altitudes and orthocenter of $\triangle ABC$

out that H does not necessarily lie inside the triangle: If $\triangle ABC$ is a right triangle, then H is at a vertex, and if $\triangle ABC$ is obtuse, then H will lie outside.

THEOREM. If a triangle $\triangle ABC$ has altitudes $\overline{AD}$, $\overline{BE}$, and $\overline{CF}$, then $\overleftrightarrow{AD}$, $\overleftrightarrow{BE}$, and $\overleftrightarrow{CF}$ meet at a point.

Proof. We cannot resist giving a very short, slick proof based on the material of the previous two chapters. Let $\triangle A''B''C''$ be the anticomplementary triangle of $\triangle ABC$. Since A is the midpoint of $\overline{B''C''}$ and since $\overline{AD}$ is perpendicular to $\overleftrightarrow{B''C''}$, we conclude that $\overleftrightarrow{AD}$ is the perpendicular bisector of $\overline{B''C''}$. Similarly, $\overleftrightarrow{BE}$ is the perpendicular bisector of $\overline{A''C''}$ and $\overleftrightarrow{CF}$ is the perpendicular bisector of $\overline{A''B''}$. But we know from Section 7.1 that the three perpendicular bisectors of the sides of any triangle meet in a point. Hence, $\overleftrightarrow{AD}$, $\overleftrightarrow{BE}$, and $\overleftrightarrow{CF}$ meet in a point, as claimed. $\square$

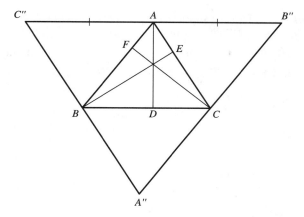

Figure 9.2: The altitudes of $\triangle ABC$ in $\triangle A''B''C''$

In addition to being short, this proof yields two corollaries. Their proofs follow from the fact that the perpendicular bisectors of the sides of a triangle meet at the circumcenter of that triangle.

> **COROLLARY.** Let $\triangle ABC$ be a triangle with anticomplementary triangle $\triangle A''B''C''$. Then the orthocenter of $\triangle ABC$ is the circumcenter of $\triangle A''B''C''$.

> **COROLLARY.** Let $\triangle ABC$ be a triangle with complementary triangle $\triangle A'B'C'$. Then the circumcenter of $\triangle ABC$ is the orthocenter of $\triangle A'B'C'$.

As an application of our theorem we can solve an "inaccessible point" construction problem. Assume that we are given two nonparallel lines ℓ_1 and ℓ_2 whose point of intersection cannot be reached. (Perhaps it is off the page.) In this problem we are given two such lines and a point P and we wish to draw the line containing P and the inaccessible intersection point of ℓ_1 and ℓ_2.

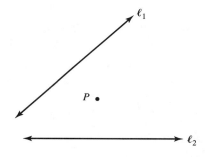

Figure 9.3: The problem

PROBLEM. Given nonperpendicular lines ℓ_1 and ℓ_2, which intersect in the inaccessible point A and given a point P, draw the line $\overleftrightarrow{AP}$.

Solution. Draw a line through P perpendicular to ℓ_1. Let this line intersect ℓ_2 at the point C. (Why must this line intersect ℓ_2?) Next, draw a line through P perpendicular to ℓ_2 intersecting the line ℓ_1 at the point B. Construct $\overleftrightarrow{BC}$. Then the line through P perpendicular to $\overleftrightarrow{BC}$ will contain A.

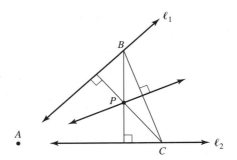

Figure 9.4: Construction of $\ell = \overleftrightarrow{PA}$

Proof. In the triangle $\triangle ABC$ the point P is the orthocenter because it is the intersection of the altitudes from B and C. Hence, the line through it and perpendicular to $\overline{BC}$ must be the altitude through A.   □

9.2 FAGNANO'S PROBLEM

In this section we will solve a geometric optimization problem due to G. C. Fagnano in 1775. Fagnano also solved it, so perhaps the result should be called Fagnano's theorem. The proof we shall give is due to L. Fejer in 1900, and our treatment closely follows that in Geometric Inequalities, by Nicholas D. Kazarinoff, in the Mathematical Association of America's New Mathematical Library, volume 4, 1961. Here is the problem.

PROBLEM. Given an acute triangle $\triangle ABC$, find points X on $\overline{BC}$, Y on $\overline{AC}$, and Z on $\overline{AB}$ that minimize the perimeter of the resulting triangle $\triangle XYZ$.

In the previous section we defined the points D, E, and F to be the points where the altitudes from A, B, and C met the opposite sides. The triangle $\triangle DEF$ is called the *orthic triangle*, of $\triangle ABC$ and we will show in this section that it is the solution to Fagnano's problem. Our proof will be based on first solving a related problem.

RELATED PROBLEM. Given an acute angled triangle $\triangle ABC$ and given a fixed point X on $\overline{BC}$, find points Y on $\overline{AC}$ and Z on $\overline{AB}$ that minimize the perimeter of $\triangle XYZ$.

(Make sure you understand the difference between the two problems. In the original problem you choose three points and in the new problem one point is given and you choose the other two.)

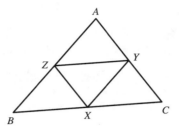

Figure 9.5: A triangle inscribed in $\triangle ABC$

Solution to the related problem. Let X' be the point such that $\overleftrightarrow{AC}$ is the perpendicular bisector of $\overline{XX'}$. Then, as in Fig. 9.6(a), if Y is any point on $\overleftrightarrow{AC}$, $\overline{XY} \cong \overline{X'Y}$. Likewise, we let X'' be the point such that $\overleftrightarrow{AB}$ is the perpendic-

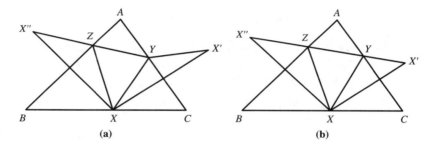

Figure 9.6: Solution of related problem

ular bisector of $\overline{XX''}$, so that if Z is any point on $\overleftrightarrow{AB}$, $\overline{XZ} \cong \overline{X''Z}$. Hence, for any choice of Y on $\overline{AC}$ and Z on $\overline{AB}$, we have the perimeter of $\triangle XYZ = XY + YZ + XZ = X'Y + YZ + ZX''$.

But, $X'Y + YZ + ZX''$ is the length of the path $X'YZX''$ from X' to X''. Therefore, to minimize the perimeter of $\triangle XYZ$, we must minimize the length of the path $X'YZX''$ joining X' to X''. But the shortest path between any two points in the plane is a straight line [cf. Fig. 9.6(b)] and this points the way to a solution of our new problem.

Given $\triangle ABC$ and a point X on $\overline{BC}$, construct X' by reflecting X through $\overleftrightarrow{AC}$ and X'' by reflecting X through $\overleftrightarrow{AB}$. Draw $\overleftrightarrow{X'X''}$ and let it intersect $\overleftrightarrow{AC}$ at Y and $\overleftrightarrow{AB}$ at Z. Then $\triangle XYZ$ is the solution to the related problem. $\square$

Solution to Fagnano's problem. In the solution to the related problem we showed how to choose Y and Z if we are given X. Our next goal is to determine a choice of X. As before, let X be on $\overline{BC}$, X' be the reflection of X through $\overleftrightarrow{AC}$, and X''

the reflection of X through $\overleftrightarrow{AB}$. We have shown that the minimum perimeter of a triangle $\triangle XYZ$ inscribed in $\triangle ABC$ with X as a vertex is the length of $\overline{X'X''}$. What we have to do now is find that particular X that minimizes the length $\overline{X'X''}$.

Since A is on the perpendicular bisector of $\overline{XX'}$, $\overline{AX} \cong \overline{AX'}$, and since A is on the perpendicular bisector of $\overline{XX''}$, $\overline{AX} \cong \overline{AX''}$, for any choice of X on $\overline{BC}$. Also, using the fact that $\triangle AXX'$ is isosceles and has congruent base angles,

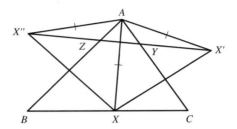

Figure 9.7: Solution of Fagnano's problem

$\angle X'AC \cong \angle CAX$; and since $\triangle AXX''$ is isosceles, we have $\angle X''AB \cong \angle BAX$. Adding, we see that $\angle X''AX' = \angle X''AB + \angle BAX + \angle XAC + \angle CAX' = 2\angle BAX + 2\angle XAC = 2\angle BAC$.

Now consider $\triangle AX'X''$. This triangle has three properties

1. $\angle X'AX'' = 2\angle BAC$, a relation that does not depend upon the choice of X.
2. The sides $\overline{AX'}$ and $\overline{AX''}$ are congruent to each other and to $\overline{AX}$.
3. The length of the base is the perimeter of $\triangle XYZ$, which we want to minimize.

Since $\triangle AX'X''$ is an isosceles triangle with a fixed summit angle, we minimize the base by minimizing the length of the legs $AX' = AX''$. But this is the length AX.

Now we are as good as done. We know that of all the line segments $\overline{AX}$ joining A to a point on $\overline{BC}$ the shortest one is the perpendicular, $\overline{AD}$. Therefore $X = D$, will be a vertex of the triangle that solves Fagnano's problem. By similar arguments the other two vertices will be E and F and the orthic triangle is the one of smallest perimeter.

We remark without proof that $\triangle DEF$ has perimeter $2K/R$, where $K = $ area and $R = $ circumradius. □

9.3 THE EULER LINE

Recall the notation $O = $ the circumcenter, $G = $ the centroid, and $H = $ the orthocenter of $\triangle ABC$. We now have the following remarkable theorem, due to Euler.

THEOREM. The three points O, G, and H lie on a straight line, with G between O and H. Moreover, $GH = 2GO$.

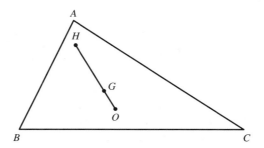

Figure 9.8: Euler line of $\triangle ABC$

Proof. As usual, we let A' be the midpoint of $\overline{BC}$ and $\overline{AD}$ the altitude to $\overline{BC}$ as in Fig. 9.9. Then $\overline{AA'}$ is a median, it contains G, and $\overleftrightarrow{A'O}$ is the perpendicular bisector of $\overline{BC}$.

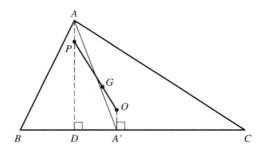

Figure 9.9: Proof that H, G, O are collinear

Connect O to G and let the resulting line intersect $\overline{AD}$ at the point P. Since $\overline{AD}$ and $\overline{OA'}$ are each perpendicular to $\overline{BC}$, they must be parallel. Hence, by the alternate interior angle theorem $\angle APG \cong \angle GOA'$ and $\angle GAP \cong \angle GA'O$. By AA, we conclude that $\triangle APG \sim \triangle A'OG$. So

$$\frac{OG}{GP} = \frac{A'G}{GA}.$$

But the centroid G trisects the median AA', so $A'G/GA = \frac{1}{2}$. Hence

$$2OG = GP.$$

This equation is the key to the proof. Consider what we just proved from the point of view of O and G. We have shown that if you extend the line segment

$\overline{OG}$ to a new point P such that G is between O and P and $GP = 2OG$, then this point P must be on the altitude $\overline{AD}$, since these two conditions determine P uniquely. However, the same manner of proof would show that this point P will be on the altitude $\overline{BE}$ and the altitude $\overline{CF}$. Thus, P is the orthocenter H and the theorem is proved.   □

The line joining O, G and H is called the *Euler line* of $\triangle ABC$.

9.4   THE NINE-POINT CIRCLE

Given a triangle $\triangle ABC$ with orthocenter H, we list nine points that will interest us (Fig. 9.10): The midpoints of the sides A', B', C'; the points of intersection of the altitudes with the opposite sides (called the feet of the altitudes) D, E, and F; and the midpoints of the segments $\overline{HA}$, $\overline{HB}$, and $\overline{HC}$. These last three points are called the Euler points and are denoted P, Q, and R. Our main theorem in this section states that all nine of these points lie on a circle. It is called the nine-point circle of $\triangle ABC$. This is remarkable in light of the fact that four arbitrary points generally do not lie on a circle, yet this theorem produces nine that do. Euler first proved in 1765 that D, E, F, A', B', and C' lie on a circle, but he missed the Euler points. Over 50 years later, C.-J. Brianchon and J.-V. Poncelet—without knowing of Euler's theorem—proved that all nine points lie on a circle.

THEOREM. The nine points A', B', C', D, E, F, P, Q, and R all lie on a circle.

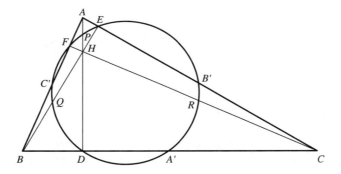

Figure 9.10: The nine-point circle

Proof. The proof will make use of three facts proven earlier. We restate them now for your convenience.

Fact 1: A line segment that joins the midpoints of two sides of a triangle is parallel to the third side and half as long (see Section 8.1).

Fact 2: In a rectangle, the two diagonals are congruent and bisect each other (see Section 2.2)

Fact 3: If a circle has diameter $\overline{AB}$ and a point C satisfies $\angle ACB = 90°$, then C is on the circle (see Section 5.2).

To begin the proof, we study the quadrilateral $C'QRB'$ (Fig. 9.11). Since C'

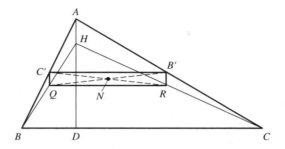

Figure 9.11: The quadrilateral $C'QRB'$

is the midpoint of $\overline{AB}$ and Q is the midpoint of $\overline{BH}$, we may apply Fact 1 to the triangle $\triangle ABH$ to see that $\overline{C'Q}$ is parallel to $\overline{AH}$ and $C'Q = \frac{1}{2}AH$. Likewise, applying Fact 1 to B' and R in $\triangle ACH$, we see that $\overline{B'R}$ is parallel to $\overline{AH}$ and $B'R = \frac{1}{2}AH$. Hence the line segments $\overline{B'R}$ and $\overline{C'Q}$ are congruent and parallel. We may conclude that $C'QRB'$ is a parallelogram.

Next, we claim that $C'QRB'$ is a rectangle. To see this, note that $\overline{C'B'}$ is parallel to $\overline{BC}$ by Fact 1, and $\overline{C'Q}$ and $\overline{B'R}$ are parallel to $\overline{AH}$, as noted previously. Since $\overline{AH}$ is perpendicular to $\overline{BC}$ it follows that $\overline{C'Q}$ and $\overline{B'R}$ are perpendicular to $\overline{C'B'}$ which shows $C'QRB'$ is a rectangle.

We are now in a position to apply Fact 2. Let N be the point of intersection of the diagonals $\overline{C'R}$ and $\overline{B'Q}$. Then Fact 2 implies that $\overline{NB'} \cong \overline{NQ} \cong \overline{NC'} \cong \overline{NR}$. Hence there is a circle with center N that contains the four points B', Q, C', R. Or, stated differently, the circle with diameter $\overline{C'R}$ contains the points B' and Q.

Now we apply the same basic argument to the quadrilateral $A'RPC'$ (Fig. 9.12). What is the conclusion? In this case we conclude that the circle with diameter $\overline{C'R}$ contains the points R, C', A', and P. Hence, this circle contains the six points A', B', C' P, Q, R, and the three line segments $\overline{A'P}$, $\overline{B'Q}$, and $\overline{C'R}$ are all diameters. To prove that the remaining three points D, E, and F also lie on this circle, we use Fact 3. Just observe that the three angles $\angle PDA'$, $\angle QEB'$, and $\angle RFC'$ are all $90°$ angles and the theorem follows.   □

We conclude this section with two additional facts about the nine-point circle.

THEOREM. The radius of the nine-point circle is one-half the circumradius of $\triangle ABC$ and the center N lies on the Euler line at the midpoint of $\overline{OH}$.

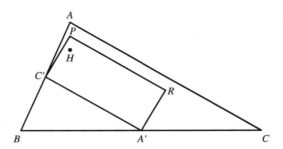

Figure 9.12: The quadrilateral $A'RPC'$

Proof. Consider the complementary triangle $\triangle A'B'C'$. Since $\triangle A'B'C'$ is similar to $\triangle ABC$ and half as big, the circumradius of $\triangle A'B'C'$ is one-half the circumradius of $\triangle ABC$. (See Exercise 1 of Chapter 7.) But the nine-point circle is the circumcircle of $\triangle A'B'C'$; thus its radius is one-half the circumradius.

Next, consider the Euler line of $\triangle A'B'C'$. The centroid of $\triangle A'B'C'$ is G, the centroid of the original triangle $\triangle ABC$ (recall Section 8.3). The orthocenter of $\triangle A'B'C'$ is O, the circumcenter of $\triangle ABC$ (recall Section 9.1). Hence, by the previous section, N lies on the line $\overleftrightarrow{OG}$, the Euler line of $\triangle ABC$.

Again, applying the results of the previous section to $\triangle A'B'C'$, the centroid G of $\triangle A'B'C'$ is between the circumcenter N of $\triangle A'B'C'$ and the orthocenter O of $\triangle A'B'C'$, closer to the circumcenter. Hence,

$$2GN = GO.$$

We leave it as an exercise for you to combine this with the equation $2GO = GH$ to prove that $ON = HN$ and N is therefore the midpoint of $\overline{OH}$.   □

From the proof we also see that the next corollary is true.

COROLLARY. A triangle and its complementary triangle have the same Euler line.

PROBLEMS

9.1. What is the orthocenter of $\triangle BCH$? (Refer to Fig. 9.1.)

*9.2. Prove that there is a circle containing the four points A, F, H, E. Prove that $BF \cdot FA = BH \cdot HE$.

9.3. (a) Prove that there is a circle containing the four points B, C, E, F.

   *(b) Prove that $\angle FEC = 180° - \angle ABC$.

   *(c) Prove that $\triangle ABC \sim \triangle AEF \sim \triangle DEC \sim \triangle DBE$

   (d) Prove that $\angle ADF \cong \angle ADE$

   (e) What is the incenter of $\triangle DEF$?

   (d)   Prove that $\angle ADF \cong \angle ADE$

   (e)   What is the incenter of $\triangle DEF$?

9.4. Our construction of a line ℓ through P and the inaccessible intersection point of ℓ_1 and ℓ_2 will not work if $\ell_1 \perp \ell_2$. Why not? Develop a construction that will work in this case.

9.5. Let $\triangle ABC$ be a right triangle with $\angle C$ the right angle. How could you choose points X on $\overline{BC}$, Y on $\overline{AC}$ and Z on $\overline{AB}$ to minimize the sum $XY + YZ + XZ$?

*9.6. Given a line ℓ and points A and B on the same side of ℓ. How could you choose a point C on ℓ to minimize the sum $AC + BC$? This problem was first solved by Heron.

9.7. Again, let $\triangle ABC$ have a right angle at $\angle C$. Prove that the median $\overleftrightarrow{CC'}$ is the Euler line. Is there any other type of triangle in which $\overleftrightarrow{CC'}$ will be the Euler line?

9.8. Prove that $\overline{OA'} \cong \overline{AP}$. (Refer to Fig. 9.9.)

9.9. Let $\triangle A'' B'' C''$ be the anticomplementary triangle of $\triangle ABC$. Identify the

   (a)   centroid

   (b)   circumcenter

   (c)   center

of the nine-point circle of $\triangle A'' B'' C''$.

9.10. Assume that we have a number line on which the circumcenter of $\triangle ABC$ is at the origin and the centroid is at the point 1. What will be the coordinates of each of each of these points?

   (a) the orthocenter

   (b) the center of the nine-point circle

   (c) the orthocenter of the complementary triangle

   (d) the orthocenter of the anticomplementary triangle

9.11. What is the nine-point circle of $\triangle HBC$?

9.12. Prove that $\triangle ABC \sim \triangle PQR$ with ratio $\frac{1}{2}$.

9.13. Let $\triangle ABC$ be equilateral.

   (a)   Prove that the nine-point circle is the same as the incircle, and conversely, that if the incircle is the same as the nine-point circle then $\triangle ABC$ is equilateral.

   (b)   Prove that the orthic triangle is the same as the medial triangle.

   (c)   Calculate the perimeter of $\triangle DEF$.

CHAPTER SUMMARY

- H is the orthocenter of $\triangle ABC$, that is, the intersection point of the altitudes drawn from the vertices to the opposite sides.

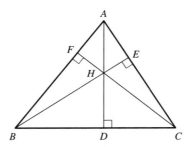

- Let $\triangle ABC$ be a triangle with anticomplementary triangle $\triangle A''B''C''$. The orthocenter of $\triangle ABC$ is the circumcenter of $\triangle A''B''C''$.

- Let $\triangle ABC$ be a triangle with complementary triangle $\triangle A'B'C'$. The circumcenter of $\triangle ABC$ is the orthocenter of $\triangle A'B'C'$.

- "Inaccessible point" construction:

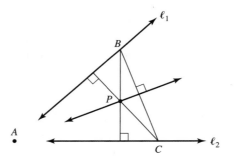

Given a point P and lines ℓ_1 and ℓ_2 that intersect in an "inaccessible" point A, the line $\overleftrightarrow{AP}$ can be drawn by first drawing a line through P perpendicular to ℓ_1 that intersects ℓ_2 at C. Next draw a line through P perpendicular to ℓ_2 intersecting ℓ_1 at B. Construct $\overline{BC}$. The line perpendicular to $\overline{BC}$ at P will contain A.

- Given an acute triangle, $\triangle ABC$, the points D, E, F (the points where the altitudes from A, B, and C meet the opposite sides) are the points on the sides of $\triangle ABC$ such that the resulting triangle, $\triangle DEF$ (called the orthic triangle), has minimum perimeter. Also, $\triangle DEF$ has perimeter $= \frac{2K}{R}$, where K is the area of $\triangle ABC$ and R is the circumradius of $\triangle ABC$.

- The three points O (circumcenter), H (orthocenter), and G (centroid) all lie on a straight line with G between H and O and with $GH = 2GO$. This line is known as the Euler line.

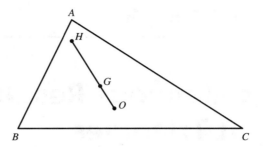

- The nine points A', B', C' (the midpoints of the sides of $\triangle ABC$), D, E, F (the feet of the altitudes of $\triangle ABC$), and P, Q, R (the Euler points, that is, P is the midpoint of $\overline{AH}$, Q is the midpoint of $\overline{BH}$, and R is the midpoint of $\overline{CH}$), all lie on a circle called the nine-point circle of $\triangle ABC$.

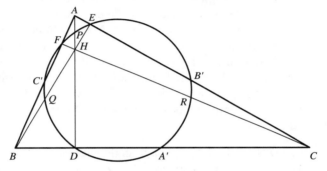

Moreover, the radius of the nine-point circle $= \frac{1}{2}$ times the circumradius, R, of $\triangle ABC$, and the center N lies on the Euler line at the midpoint of $\overline{OH}$, with $2GN = GO$ (See the figure). Also, a triangle, $\triangle ABC$, and its complementary triangle, $\triangle A'B'C'$, have the same Euler line.

C H A P T E R   10

Miscellaneous Results about Triangles

10.1 CEVA'S THEOREM

Up until now we have proven a number of theorems that stated that three lines associated with a triangle meet in one point. In each of the cases (angle bisectors, medians, altitudes, etc.) the proofs were different. You might be wondering if there isn't some "master theorem" in this subject, some grand theorem that tells when three such lines meet in a point from which we could derive all of our theorems as special cases. In this section we will prove a general theorem about lines in a triangle meeting in a point. It will not cover quite all of the cases we have studied, and it may not always give an easier or more intuitive proof than we could get without it, but it is a useful and interesting result.

THEOREM. If $\triangle ABC$ is a triangle with points X on $\overline{BC}$, Y on $\overline{AC}$, and Z on $\overline{AB}$, then the three segments $\overline{AX}$, $\overline{BY}$, and $\overline{CZ}$ meet in one point if and only if the following equation holds:

$$\frac{BX}{CX} \cdot \frac{CY}{AY} \cdot \frac{AZ}{BZ} = 1.$$

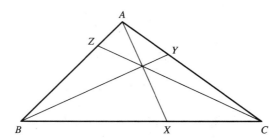

Figure 10.1: Three cevians that meet in a point

This theorem is due to Giovanni Ceva in 1678. In honor of it, a line joining the vertex of a triangle to a point on the opposite side is called a cevian. You may wish to look ahead at the next section to see how this theorem is applied before reading the proof.

Proof. Since Ceva's theorem is an "if and only if" theorem, we have two statements to prove. First we will assume that $\overline{AX}$, $\overline{BY}$, and $\overline{CZ}$ meet in a point P and prove that the above equation holds true. The proof will be based on a consideration of areas in Fig. 10.2. First, we let $\overline{AD}$ be an altitude and compare the

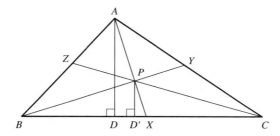

Figure 10.2: Proof of Ceva's theorem

areas of triangles $\triangle ABX$ and $\triangle ACX$. Each has height AD and the two bases are BX and CX, respectively. Hence,

$$\frac{\text{area } \triangle ABX}{\text{area } \triangle ACX} = \frac{\frac{1}{2} \cdot AD \cdot BX}{\frac{1}{2} \cdot AD \cdot CX} = \frac{BX}{CX}.$$

The point is that if two triangles have the same vertex and bases along the same line, then the ratio of their areas equals the ratio of their bases.

Now, apply the same ideas to $\triangle PBX$ and $\triangle PXC$. If we let the height be PD' we see that

$$\frac{\text{area } \triangle PBX}{\text{area } \triangle PCX} = \frac{\frac{1}{2} \cdot PD' \cdot BX}{\frac{1}{2} \cdot PD' \cdot CX} = \frac{BX}{CX}.$$

We now cross-multiply in each of our two equations, which yields
$$CX \cdot \text{area } \triangle ABX = BX \cdot \text{area } \triangle ACX$$

and
$$CX \cdot \text{area } \triangle PBX = BX \cdot \text{area } \triangle PCX.$$
Subtracting the bottom equation from the top one, we see that
$$CX \, (\text{area } \triangle ABX - \text{area } \triangle PBX) =$$
$$BX \, (\text{area } \triangle ACX - \text{area } \triangle PCX).$$

From the picture, area $\triangle ABX -$ area $\triangle PBX =$ area $\triangle APB$ and area $\triangle ACX -$ area $\triangle PCX =$ area $\triangle APC$, so after substituting and forming ratios we have

$$\frac{BX}{CX} = \frac{\text{area } \triangle APB}{\text{area } \triangle APC}.$$

This is the equation we need. It gives a useful description of the ratio in which the point X divides the line segment $\overline{BC}$. We repeat the process to evaluate CY/AY and AZ/BZ. We obtain

$$\frac{CY}{AY} = \frac{\text{area } \triangle BPC}{\text{area } \triangle BPA}$$

and

$$\frac{AZ}{BZ} = \frac{\text{area } \triangle CPA}{\text{area } \triangle CPB}.$$

If we now substitute these three area ratios into the expression

$$\frac{BX}{CX} \cdot \frac{CY}{AY} \cdot \frac{AZ}{BZ}$$

we see that everything cancels and we are just left with 1. This proves half of the theorem. We now turn to the converse.

We assume that

$$\frac{BX}{CX} \frac{CY}{AY} \frac{AZ}{BZ} = 1$$

and we will prove that $\overline{AX}$, $\overline{BY}$, and $\overline{CZ}$ meet in one point. Let P be the point of intersection of $\overline{BY}$ and $\overline{CZ}$. We need to show that $\overline{AX}$ also goes through the point P. Draw the line $\overleftrightarrow{AP}$ and assume that $\overleftrightarrow{AP}$ intersects $\overline{BC}$ at X' as in Fig. 10.3. We will show that $X = X'$. In fact, we've already done all of the hard

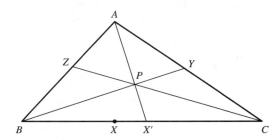

Figure 10.3: Proof of converse of Ceva's theorem

work: We know from the half of the theorem we've already proved that

$$\frac{BX'}{CX'} \cdot \frac{CY}{AY} \cdot \frac{AZ}{BZ} = 1.$$

Comparing this to our hypothesis that

$$\frac{BX}{CX} \cdot \frac{CY}{AY} \cdot \frac{AZ}{BZ} = 1,$$

we see that

$$\frac{BX'}{CX'} = \frac{BX}{CX}.$$

Now we are done, since there is only one point betweeen B and C that divides it in this ratio, for if X' were to the right of X, then $\frac{BX'}{CX'}$ would be greater than BX/CX, and if it were to the left, $\frac{BX'}{CX'}$ would be smaller. We conclude that $X = X'$ and Ceva's theorem is proved.   □

10.2   APPLICATIONS OF CEVA'S THEOREM

As applications, we will reprove the theorems that the three medians meet in a point and that the three angle bisectors meet in a point. Additional applications may be found in the exercises.

THEOREM. The three medians $\overline{AA'}$, $\overline{BB'}$, and $\overline{CC'}$ meet in a point.

Proof. We apply Ceva's theorem with $X = A'$, $Y = B'$ and $Z = C'$. In this case $BA' = CA'$, so $BA'/CA' = 1$. Likewise, $CB'/AB' = 1$ and $AC'/BC' = 1$. Hence,

$$\frac{BX}{CX} \cdot \frac{CY}{AY} \cdot \frac{AZ}{BZ} =$$

$$\frac{BA'}{CA'} \cdot \frac{CB'}{AB'} \cdot \frac{AC'}{BC'} =$$

$$1 \cdot 1 \cdot 1 = 1.$$

Ceva's theorem immediately implies that $\overline{AA'}$, $\overline{BB'}$, and $\overline{CC'}$ meet in one point.
□

In order to use Ceva's theorem to prove that the angle bisectors meet in a point, we need to study how the point of intersection of an angle bisector with the opposite side of the triangle divides that side. The answer is provided by the next lemma (see Exercise 2 for an alternate method of proof).

LEMMA. If the bisector of $\angle BAC$ intersects $\overline{BC}$ at the point X, then

$$\frac{BX}{CX} = \frac{AB}{AC}.$$

Before we prove this lemma, we will show how to combine it with Ceva's theorem to prove that the three angle bisectors meet in a point. Let X be the point where the bisector of $\angle A$ intersects $\overline{BC}$, Y the point where the bisector of $\angle B$ intersects $\overline{AC}$, and Z the point where the bisector of $\angle C$ meets $\overline{AB}$. Also, as usual, denote the sides of the triangle by $BC = a$, $AC = b$, and $AB = c$. We need to compute

$$\frac{BX}{CX} \cdot \frac{CY}{AY} \cdot \frac{AZ}{BZ}$$

and show that it equals 1. By the lemma, $BX/CX = c/b$, $CY/AY = a/c$ and $AZ/BZ = b/a$. So the product is $\frac{c}{b} \cdot \frac{a}{c} \cdot \frac{b}{a} = 1$, as claimed. We now turn to the proof of the lemma.

Proof of Lemma: Draw a line through B (as in Fig. 10.4) parallel to $\overline{AX}$ which intersects $\overleftrightarrow{AC}$ at the point A_1.

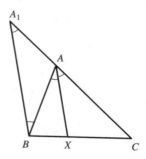

Figure 10.4: The ratio BX/CX equals c/b

Since $\overline{AX} \| \overline{A_1 B}$, $\angle CAX \cong \angle A_1$ and $\angle CXA \cong \angle CBA_1$, as corresponding angles. Hence, $\triangle CAX \sim \triangle CA_1 B$ by AA. This implies that

$$\frac{CB}{CX} = \frac{CA_1}{CA}.$$

If we subtract 1 from both sides of the preceeding equation we get

$$\frac{CB - CX}{CX} = \frac{CA_1 - CA}{CA},$$

or

$$\frac{BX}{CX} = \frac{AA_1}{CA}.$$

This is close to what we want: It involves BX, CX, and CA, but it has AA_1 instead of AB. To complete the proof of the lemma we need to prove that $\overline{AB} \cong \overline{AA_1}$.

Since $\overline{AX} \| \overline{A_1 B}$ and $\overline{AB}$ is a transversal, $\angle ABA_1 \cong \angle BAX$, for they are alternate interior angles. If we chain together the three congruences

$$\angle ABA_1 \cong \angle BAX,$$

$$\angle A_1 \cong \angle CAX \text{ (proven previously)},$$

and

$$\angle BAX \cong \angle CAX \text{ (since } AX \text{ is the bisector)},$$

it follows that $\angle A_1 \cong \angle ABA_1$. Hence $\triangle AA_1 B$ is isosceles and $\overline{AA_1} \cong \overline{AB}$, which completes the proof. $\square$

10.3   THE FERMAT POINT

Sometimes a good question may be more important than a good answer. There was a bitter dispute between Torricelli and Fermat over who first discovered this point. However, Torricelli's work came about because of a question asked by Fermat, and it is Fermat who is generally honored by having this point named after him. We saw a similar historical phenomenon in Section 9.2 in that Fagnano is referred to most often as the person who posed a problem rather than as the one who solved it, even though he also did both.

In our discussion of the Fermat point, which will include this section and the next one, we will make the simplifying assumption that $\triangle ABC$ is a triangle with all angles less than 120° i.e., $\angle A < 120°$, $\angle B < 120°$ and $\angle C < 120°$. Our basic result is as follows:

THEOREM. There is a unique point P inside of $\triangle ABC$ such that

$$\angle APB = \angle BPC = \angle CPA = 120°.$$

This point is called the *Fermat point* of $\triangle ABC$, and we will show why it is of interest in the next section.

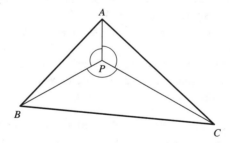

Figure 10.5: Fermat point of $\triangle ABC$

We now describe a construction which will be useful in our study of the Fermat point. Given $\triangle ABC$ we construct (as in Fig. 10.6) three equilateral triangles $\triangle BCA'$, $\triangle ACB'$, and $\triangle ABC'$, that lie outside $\triangle ABC$, with bases $\overline{BC}$, $\overline{AC}$, and $\overline{AB}$.

If we construct the circumcircles of the triangles $\triangle ACB'$ and $\triangle ABC'$, they intersect at the point A and, in general, at one other point P. Under the assumption that all of the angles of $\triangle ABC$ are less than 120°, this second point of intersection P will be inside of $\triangle ABC$. Even though we will use this fact, we will not prove it here. The proof is a bit technical, and we have left it as Exercise 5 at the end of this chapter.

We may now easily prove half of our theorem.

LEMMA. Let P be the point of intersection of the circumcircles of $\triangle ABC'$ and $\triangle ACB'$. Then $\angle APB = \angle BPC = \angle CPA = 120°$.

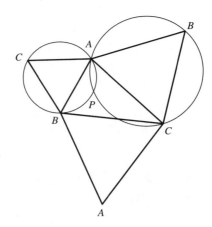

Figure 10.6: Construction of Fermat point

Proof. In the circumcircle of $\triangle ABC'$ we have $\angle APB = \frac{1}{2}\overarc{AC'B}$. But $\overarc{AC'B} = 360° - \overarc{APB}$ and $\overarc{APB} = 2\angle C' = 2 \cdot 60° = 120°$. Hence $\overarc{AC'B} = 240°$ and $\angle APB = 120°$, as claimed. A similar argument using the circumcircle of $\triangle ACB'$ shows that $\angle CPA = 120°$. Finally, $\angle BPC = 360° - \angle APB - \angle CPA = 360° - 120° - 120° = 120°$. $\qquad\square$

To complete the proof of the theorem, we need to establish the uniqueness of P: P is the only point that makes equal angles with $\overline{AB}$, $\overline{BC}$, and $\overline{AC}$. To this end we prove a lemma about equilateral triangles.

LEMMA. Let $\triangle ABC$ be an equilateral triangle and let P be a point on the opposite side of $\overleftrightarrow{BC}$ from A. Then $\angle BPC = 120°$ if and only if P is on the circumcircle of $\triangle ABC$.

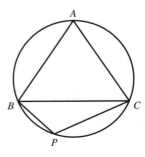

Figure 10.7: Angle P on circumcircle

Proof. First, assume that P is on the circumcircle. Then $\angle BPC = \frac{1}{2} \ \overset{\frown}{BAC} = \frac{1}{2}(360° - \overset{\frown}{BPC}) = \frac{1}{2}(360° - 2\angle ABC) = \frac{1}{2}(360° - 2 \cdot 60°) = 120°$. This is essentially the same as the proof of the previous lemma.

To prove the converse, assume that $\angle BPC = 120°$. We want to show that P is on the circumcircle. Our proof will be by contradiction. If P does not lie on the circumcircle, then $\overrightarrow{BP}$ will intersect this circle (see Fig. 10.8) at some other point P'. Consider the triangle $\triangle CPP'$. By assumption, $\angle CPB = 120°$ and, by

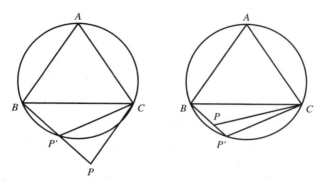

Figure 10.8: $\angle P = 120°$, P outside and inside the circle

the preceeding proof, $\angle CP'B = 120°$. But this situation is impossible, since an exterior angle of a triangle cannot equal an opposite interior angle.   □

We may now proceed to establish the uniqueness of P.

Proof of theorem: We have already shown that there is a point P in $\triangle ABC$ that subtends 120° angles with each of the three sides. We now need to show that P is the only point with this property. So assume that Q is another point inside $\triangle ABC$ such that $\angle AQB = \angle BQC = \angle CQA = 120°$. Since $\angle AQB = 120°$, the last lemma says Q must lie on the circumcircle of $\triangle ABC'$; similarly, since $\angle AQC = 120°$, Q must lie on the circumcircle of $\triangle ACB'$. Consequently, Q must lie on the intersection of these two circles forcing $P = Q$.   □

Note that since $\angle BPC = 120°$, the Fermat point also lies on the circumcircle of $\triangle BCA'$. In particular, these three circles meet at a point, which we note as the next result.

COROLLARY. The circumcircles of the three triangles $\triangle BCA'$, $\triangle ACB'$, and $\triangle ABC'$ meet in a point.

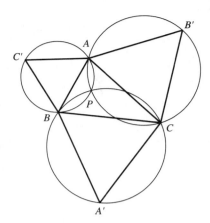

Figure 10.9: Three circumcircles meet at P, the Fermat point.

10.4 PROPERTIES OF THE FERMAT POINT

Fermat posed the following geometric optimization problem to Torricelli: Given a triangle $\triangle ABC$, find a point P that minimizes the sum $PA + PB + PC$. This is related to the practical problem of constructing a shortest possible network of, say, cables to connect a number of cities. Torricelli's solution was the point we now call the Fermat point of $\triangle ABC$. Recall that we are assuming that all angles of $\triangle ABC$ are less than $120°$.

THEOREM. Of all points in $\triangle ABC$, the Fermat point has the minimal sum of distances to the three vertices.

Proof. Let Q be any point inside $\triangle ABC$. We will study the sum $QA + QB + QC$.

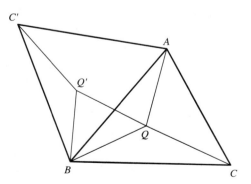

Figure 10.10: Construction of Q' and C' from Q

Construct the point C' as in Section 10.3, making $\triangle ABC'$ an equilateral triangle. Now, using $\overline{C'B}$ as the base, construct a triangle $\triangle C'BQ'$ congruent to $\triangle ABQ$, as shown in Fig. 10.10. Since these two triangles are congruent, $\overline{C'Q'} \cong \overline{AQ}$ and $\overline{BQ'} \cong \overline{BQ}$.

Next we claim that $\triangle BQQ'$ is equilateral. Since $\overline{BQ'} \cong \overline{BQ}$, it must be isosceles. To prove that it is actually equilateral, we prove that $\angle QBQ'$ is a 60° angle. First observe that

$$\angle QBQ' = \angle CBC' - \angle C'BQ' - \angle CBQ$$

$$= \angle CBC' - (\angle C'BQ' + \angle CBQ).$$

Now consider each of the terms in this last expression. We have

$$\angle CBC' = \angle CBA + \angle ABC' = \angle CBA + 60°.$$

Also, as corresponding parts of congruent triangles, $\angle C'BQ' \cong \angle ABQ$ and therefore

$$\angle C'BQ' + \angle CBQ = \angle ABQ + \angle CBQ = \angle CBA.$$

Substituting all of this into the expression for $\angle QBQ'$ yields

$$\angle QBQ' = \angle CBC' - (\angle C'BQ' + \angle CBQ)$$

$$= 60° + \angle CBA - \angle CBA$$

$$= 60°, \text{ as claimed.}$$

So $\triangle BQQ'$ is equilateral.

This is important because it implies that $\overline{QQ'} \cong \overline{BQ}$. Because we have also constructed Q' so that $\overline{C'Q'} \cong \overline{AQ}$, we may represent our sum as $AQ + BQ + CQ = C'Q' + Q'Q + QC$. This says that for any point Q in $\triangle ABC$, the sum in which we are interested equals the length of the path $C'Q'QC$ from C' to C.

If we could somehow choose Q so that Q and Q' lie on the line segment $\overline{CC'}$, then this choice of Q would give the shortest possible path from C to C' and would also minimize the sum $AQ + BQ + CQ$. It turns out that if $Q = P$, the Fermat point, then the path $C'Q'QC$ will be a straight line.

To show this we need to show that $\angle CQQ' = 180°$ and $\angle QQ'C' = 180°$ when $Q = P$. But, $\angle CQQ' = \angle CQB + \angle BQ'Q$. However, $\angle CQB = 120°$, by the definition of the Fermat point, and $\angle BQ'Q = 60°$ since $\triangle BQQ'$ is equilateral. So $\angle CQQ' = 120° + 60° = 180°$. Likewise, $\angle QQ'C' = \angle BQ'Q + \angle BQ'C'$. Here, $\angle BQ'Q = 60°$ and $\angle BQ'C' = 120°$ since $\triangle BQ'C' \cong \triangle BQA$. This shows that if $Q = P$, then the path $C'Q'QC$ is as short as possible, so the sum $QA + QB + QC$ is small as possible, or, in other words, minimal. This proves the theorem. □

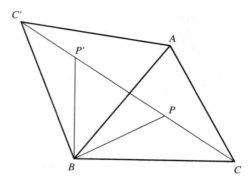

Figure 10.11: Proof that $C'P'PC$ is a straight line

It follows from the proof that P lies on the line segment $\overline{CC'}$ and that the length of $\overline{CC'}$ is $PA + PB + PC$. By replacing C in this last sentence by A or B, we obtain a proof of the following amazing result.

COROLLARY. (1) $\overline{AA'} \cong \overline{BB'} \cong \overline{CC'}$.
  (2) $\overline{AA'}$, $\overline{BB'}$, and $\overline{CC'}$ meet at P.

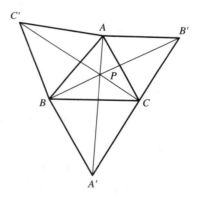

Figure 10.12: $\overline{AA'}$, $\overline{BB'}$ and $\overline{CC'}$ meet at P

Note that any point Q that minimizes the sum $QA + QB + QC$ must satisfy (2), which implies that P is the unique such point. We also remark that it seems to us it would be quite hard to prove (2) using Ceva's theorem.

There are a number of sources of additional information on Fermat's theorem. You may enjoy reading Chapter 2 of *Mathematical Gems* by Ross Honsberger, in the Dolciani Mathematical Expositions series of the Mathematical Association of America, 1973. Our exposition is based on Honsberger's; "The generalized Fermat Point," by J. Tong and Y. S. Chua in *Mathematics Magazine,* 68(1995), pp. 214–215 or *A Generalization of the Fermat-Torricelli Point*, by M.D. deVilliers, in *Mathematical Gazette,*

alization of the Fermat-Torricelli Point, by M.D. deVilliers, in *Mathematical Gazette,* 79(1995), pp. 374–8.

PROBLEMS

*10.1. Use Ceva's theorem to prove that the altitudes of an acute triangle meet in a point.

10.2. Let the bisector of $\angle BAC$ intersect $\overline{BC}$ at X. Prove that $BX/BC = c/(b+c)$ by comparing the areas of $\triangle ABX$ and $\triangle ACX$ two different ways: Use the fact that they have equal altitudes and the fact that they have a congruent angle.

*10.3. Let the inscribed circle of $\triangle ABC$ be tangent to $\overline{BC}$ at the point X, $\overline{AC}$ at the point Y and $\overline{AB}$ at the point Z. Then prove that $\overline{AX}$, $\overline{BY}$, and $\overline{CZ}$ meet at a point.

This point of intersection is called the *Georgonne* point of $\triangle ABC$.

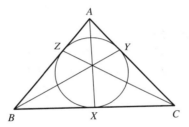

Figure 10.13: Exercise 3

10.4. Let the midpoints of the sides of $\triangle ABC$ be A', B', and C' as usual. Let X, X' be points on $\overline{BC}$; Y, Y' be points on $\overline{AC}$; and Z, Z' be points on $\overline{AB}$ such that A' is the midpoint of $\overline{XX'}$, B' is the midpoint of $\overline{YY'}$, and C' is the midpoint of $\overline{ZZ'}$. Prove that $\overline{AX}$, $\overline{BY}$, and $\overline{CZ}$ meet at a point if and only if $\overline{AX'}$, $\overline{BY'}$, and $\overline{CZ'}$ meet at a point.

10.5. Let P be the intersection point of the circumcircles of $\triangle ABC'$ and $\triangle ACB'$.
  *(a) Prove that if P lies in region 1, then $\angle A$ must be greater than 120°
  *(b) If P is in region 2, then $\angle B$ must be greater than 120° (see Fig. 10.14).

10.6. Let $\triangle ABC$ be any triangle and let D be on the opposite side of $\overrightarrow{BC}$ from A. Prove that D is on the circumcircle of $\triangle ABC$ if and only if $\angle BDC = 180° - \angle A$. Use this fact to prove that a quadrilateral $ABCD$ can be inscribed in a circle if and only if $\angle A + \angle D = 180°$.

10.7. Prove that $\triangle AC'C \cong \triangle AB'B$, in the figure for Exercise 5.

10.8. If $\angle A = 120°$ prove that $A = P$, in the figure for Exercise 5.

10.9. What is the solution to Fermat's problem if $\angle A$ is greater than 120°?

10.10. Given equilateral triangle, $\triangle ABC$, calculate $PA + PB + PC$.

*10.11. Given a line segment $\overline{AB}$ and a line ℓ that doesn't intersect it, find a point C on ℓ to minimize the sum $AC + BC$.

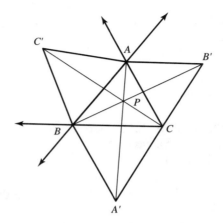

Figure 10.14: Exercise 5

inscribed quadrilateral $EFGH$.

10.13. Given a quadrilateral $ABCD$ and points E on $\overline{AB}$ and F on $\overline{BC}$, let E' be the projection of E through $\overleftrightarrow{AD}$ and let F' be the projection of F through $\overleftrightarrow{DC}$. Assume that $\overline{E'F'}$ intersects $\overline{CD}$ at G and $\overline{AD}$ at H. Prove that $EFGH$ has minimal perimeter of all inscribed quadrilaterals with vertices E and F.

CHAPTER SUMMARY

- Given a triangle, $\triangle ABC$, with points X on $\overline{BC}$, Y on $\overline{AC}$, and Z on $\overline{AB}$, the three lines (called cevians) $\overline{AX}$, $\overline{BY}$, and $\overline{CZ}$ meet in a point if and only if $\dfrac{BX}{CX} \cdot \dfrac{CY}{AY} \cdot \dfrac{AZ}{BZ} = 1$.

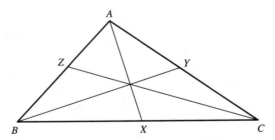

- The facts that the three medians meet in a point and that the three angle bisectors meet in a point can be reproven using Ceva's theorem.

- There is a unique point P inside of $\triangle ABC$ such that $\angle APB \cong \angle BPC \cong \angle CPA = 120°$. This point is called the Fermat point.

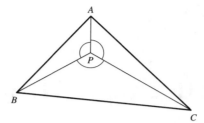

- The Fermat point, P, is the point of intersection of the circumcircles of $\triangle ABC'$, $\triangle ACB'$, and $\triangle A'BC$, the equilateral triangles constructed with bases $\overline{AB}$, $\overline{AC}$ and $\overline{BC}$, respectively. Of all the points in $\triangle ABC$, the Fermat point minimizes the sum of the distances to the three vertices, $PA + PB + PC$.

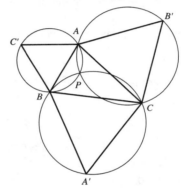

- Also, (refer to figure below)

  1. $\overline{AA'} \cong \overline{BB'} \cong \overline{CC'} = PA + PB + PC$
  2. $\overline{AA'}$, $\overline{BB'}$ and $\overline{CC'}$ meet at P.

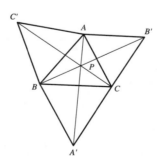

CHAPTER 11

Constructions with Indirect Elements

11.1 CONSTRUCTIONS WITH INDIRECT ELEMENTS

After sidelining construction problems for several chapters, we now conclude with a short chapter on constructions. We will be interested in situations in which we are given an unknown triangle and various clues as to what it looks like. We would like to construct a triangle $\triangle ABC$ congruent to it. These clues might include lengths of sides a, b, c; sizes of angles $\angle A$, $\angle B$, $\angle C$; lengths of medians m_a, m_b, m_c; altitudes h_a, h_b, h_c; circumradius R; or various expressions involving these. We jump right in with some examples.

EXAMPLE 1. Construct $\triangle ABC$ given a, m_a, and h_a.

Solution. We will use the three clues one at a time. First construct a line segment of length a. We may label this $\overline{BC}$ and take it to be the base of our triangle. All that remains is to find a point A that gives $\triangle ABC$ with median and altitude m_a and h_a from A to $\overline{BC}$. Since we have $\overline{BC}$ we may find the midpoint A' and draw a circle with center A' and radius m_a. (Fig. 11.1). Since AA' is to be m_a,

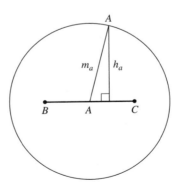

Figure 11.1: Example 1

the vertex A must be on this circle. We now use the altitude h_a to complete our search for A. Draw a line parallel to $\overline{BC}$ and at a distance h_a from it. (One way to do this would be to draw a line segment of length h_a perpendicular to $\overline{BC}$ and a line through the end point of the segment and perpendicular to it.) Since the distance from A to $\overline{BC}$ is h_a, A must lie on this line. Hence, we may take A to be either one of the two points where the line and circle intersect, and we are done.

EXAMPLE 2. Construct $\triangle ABC$ given a, R, and m_a.

Solution. The first step in doing these problems is to guess which of the three pieces of information to use first. In the previous example we started off with $a = BC$. This is often a good guess. In this problem, however, it is easier to start off with R, the circumradius.

Choose any point O and draw a circle with center O and radius R. Choose any point on the circle and label it B. Find a point C on the circle such that $\overline{BC}$ has length a. (C would be a point of intersection of the circumcircle with a circle centered at B with radius length a.) Finally, we utilize m_a as in Example 1: We find the midpoint A' of $\overline{BC}$ and construct a circle with center at A' and radius m_a. Then A will be the intersection point of this circle with the circumcircle.

EXAMPLE 3. Construct $\triangle ABC$ given $\angle A$, h_b, and h_c.

Solution. Let A be any point and construct $\angle PAQ \cong \angle A$. B will be some point on $\overrightarrow{AP}$ and C will be some point on $\overrightarrow{AQ}$. How can we identify them? Let ℓ_1 be a line on the P side and parallel to $\overleftrightarrow{AQ}$ and at a distance h_b from it. We want B on this line, so take B to be the intersection point of ℓ_1 and $\overrightarrow{AP}$. Likewise, if we construct ℓ_2 on the Q side parallel to $\overleftrightarrow{AP}$ and at a distance h_c, then C will be the intersection point of ℓ_2 with $\overrightarrow{AQ}$.

EXAMPLE 4. Construct $\triangle ABC$ given $\angle A$, h_a, and b.

Solution. Construct a line segment of length b and label it $\overline{AC}$. Draw a ray $\overrightarrow{AP}$ such that $\angle PAC \cong \angle A$. B will be some point on $\overrightarrow{AP}$ (Fig. 11.2). To find which one we will use h_a. Draw a circle with center A and radius h_a. Draw the line through C and tangent to this circle at D. Extend $\overline{CD}$ to intersect $\overrightarrow{AP}$ at B. Then $\triangle ABC$ is the desired triangle, because $AD = h_a$ and $\overline{AD}$ is perpendicular to $\overline{BC}$ since a radius of a circle drawn to a point of tangency is perpendicular to the tangent.

EXAMPLE 5. Construct $\triangle ABC$ given a, m_b, and m_c.

Solution. In order to do this problem it is necessary to use the theory of medians from Chapter 8, especially Section 8.1. The important fact is that the medians intersect at a point G which trisects them.

As usual, we start off by constructing a line segment of length a and labelling it $\overline{BC}$. Now, instead of trying to get A directly, we first construct G. Since

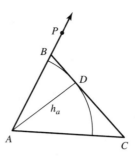

Figure 11.2: Example 4

$GB = \frac{2}{3}m_b$ and $GC = \frac{2}{3}m_c$, we may construct G (Section 4.2). Now connect G to A', the midpoint of $\overline{BC}$¡ and extend $\overline{GA'}$ to the point A such that $2AG = GA'$. Then $\triangle ABC$ is the required triangle.

EXAMPLE 6. Construct $\triangle ABC$ given $\angle A$, a, and h_a.

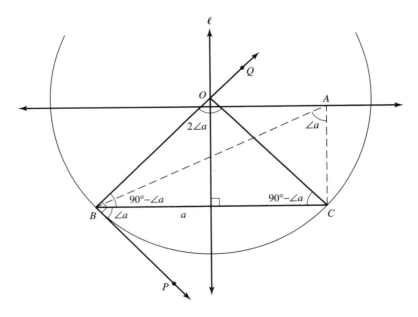

Figure 11.3: Example 6

Solution. Construct $\overline{BC}$ to have length a. Take ℓ as the perpendicular bisector of $\overline{BC}$. Construct $\angle CBP$ to have measure of $\angle A$ as shown. Next, construct $\angle PBQ$ to have a measure of $90°$. Thus, $\angle QBC$ has measure $90° - \angle A$. Let O be the intersection of $\overline{BQ}$ and ℓ. Now $\angle BOC = 2\angle A$. Draw the circle with center

O and radius $\overline{OC}$. Choose point A on this circle so that the distance of A from $\overline{BC}$ is h_a. (Just construct a line parallel to $\overline{BC}$ a distance of h_a from $\overline{BC}$ and see where it intersects the circle.) But $\angle BAC$ intercepts the same arc as $\angle CBP$, therefore $\triangle ABC$ is the desired triangle.

As a sort of dessert after the meal, we conclude this chapter and our study of Euclidean plane geometry with a long list of construction problems. Have fun!

PROBLEMS

Construct $\triangle ABC$ given the following:

11.1. a, b, c

11.2. A, b, c

11.3. R, b, c

11.4. $R, \angle A, b$

11.5. $a, m_a, K (= \text{area})$

11.6. a, R, h_a

11.7. $a, \angle B, m_a$

11.8. $\angle B, h_a, m_a$

11.9. $\angle B, a, m_c$

11.10. b, c, h_a

11.11. $\angle A, a, h_b$

*11.12. $\angle B, \angle C, h_a$

*11.13. $\angle A$, the length of the bisector of $\angle A$, r

*11.14. $b^2 + c^2, a, h_a$

*11.15. m_a, m_b, m_c

C H A P T E R   12

Solid Geometry

12.1 LINES AND PLANES IN SPACE

Whereas our approach in this chapter will be the same as in plane geometry parts of the book (stressing results rather than axiomatics), we will start with three axioms that govern relations among points, lines, and planes. First, we want to reassure you that most of the material learned in earlier chapters remains correct when lines, triangles, and associated circles are considered in space rather than as merely figures in a plane. In particular, all the results concerning congruence and similarity of triangles remain just as valid in space as they were in the plane. On the other hand, the plane geometry result that it is possible to construct one and only one perpendicular to a given line at a given point upon it is no longer true in space. This is one example of a plane geometry fact that must be modified to produce a correct solid geometry result:

Here are the three axioms we need.

(S1) There is a unique plane containing three given noncollinear points. If A, B, and C are the given points, then we will denote the one plane they determine by $\pi(A, B, C)$.

(S2) If two distinct points are in a plane, then the entire line they determine also is in the plane.

(S3) If two planes intersect, then their intersection consists of more than one point.

We gather some immediate consequences of these assumptions into our first lemma.

LEMMA. (1) If ℓ is a line and P is a point not on ℓ, then there is a unique plane π that contains P and ℓ. [We will sometimes denote π by $\pi(P, \ell)$.]

(2) If ℓ_1 and ℓ_2 are distinct intersecting lines, then there is a unique plane π which contains both ℓ_1 and ℓ_2. [We denote this plane at times by $\pi(\ell_1, \ell_2)$.]

(3) There is at most one plane containing two distinct lines.

Proof. (1) Let A and B be two distinct points (Fig. 12.1) on ℓ. By (S1), the three distinct points A, B, and P determine a unique plane $\pi(A, B, P)$. Now we set $\pi(A, B, P) = \pi$, and show that π is the plane we seek. First, from (S2), since A and B are on π, π must contain $\overleftrightarrow{AB} = \ell$. Second, suppose π' is another plane containing P and ℓ. But then π' contains P, A, and B, which, by (S1), means $\pi' = \pi(A, B, P)$. Thus we must have $\pi' = \pi$ and π is unique.

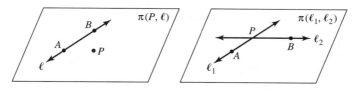

Figure 12.1: $\pi(P, \ell)$ and $\pi(\ell_1, \ell_2)$

(2) To prove part (2) of the lemma, let P be the point of intersection of ℓ_1 and ℓ_2. Moreover, suppose A is a point of ℓ_1 not on ℓ_2 and B is a point of ℓ_2 not on ℓ_1 (Fig. 12.1). We claim that $\pi = \pi(A, B, P)$ does the job. Since $\ell_1 = \overleftrightarrow{AP}$, from part (1) of this lemma, $\pi = \pi(B, \ell_1)$ and, since $\ell_2 = \overleftrightarrow{BP}$, we similarly have $\pi = \pi(A, \ell_2)$. Finally, any plane containing ℓ_1 and ℓ_2 must contain the three points A, B, and P, and thus must coincide with π.

(3) If the two lines intersect, then the plane of part (2) is the one we want. If the two lines do not intersect, then they either are parallel lines in a plane or they are not. Parallel lines in a plane determine this plane uniquely by part (1) and if the lines neither intersect nor are parallel in a plane, then no plane contains them. This proves the lemma. □

We refer to distinct lines of the type occuring in part (3) of the lemma that do not intersect and are contained in no plane as skew lines. Naturally, nonintersecting lines in a plane are still called parallel. We are now in a position to prove our first basic theorem.

THEOREM. If two distinct planes intersect, then their intersection is a line.

Proof. Let π_1 and π_2 be the two distinct intersecting planes, and from (S3) we know there must be distinct points P and Q in their intersection. We assert that the intersection of π_1 and π_2 is equal to the line $\overleftrightarrow{PQ}$. Because P and Q are on π_1, it must be by (S2) that π_1 contains $\overleftrightarrow{PQ}$ and, likewise, because P and Q are on π_2, π_2 contains $\overleftrightarrow{PQ}$. Thus the intersection of π_1 and π_2 contains $\overleftrightarrow{PQ}$. Now, suppose we could find a point A in the intersection of π_1 and π_2 that is not on $\overleftrightarrow{PQ}$. Applying part (1) of our lemma, we see that A and $\overleftrightarrow{PQ}$ are both contained in π_1, implying that $\pi_1 = \pi(A, \overleftrightarrow{PQ})$ and, similarly, we could conclude that $\pi_2 = \pi(A, \overleftrightarrow{PQ})$. However, these equalities force $\pi_1 = \pi_2$, which contradicts their distinctness. Therefore, there is no point of intersection that is not on $\overleftrightarrow{PQ}$ and, as asserted, the planes π_1 and π_2 must meet in a line. □

The next theorem is also a basic result and offers the first challenge in spatial perception. As an aid to our visualization, we break the three-space proof into its "planar pieces" in several figures that accompany the proof.

THEOREM. Let ℓ_1 and ℓ_2 be distinct lines in the plane π that intersect at the point F. If P is a point that is not on π with $\overleftrightarrow{PF} \perp \ell_1$ and $\overleftrightarrow{PF} \perp \ell_2$, then $\overleftrightarrow{PF}$ is perpendicular to any line in π that passes through F.

Proof. Choose points A, C on ℓ_1 and B, D on ℓ_2 (as in Fig. 12.2) so that $FA = FB = FC = FD$. Let ℓ be a line in π that passes through F and that we may suppose (by relettering if necessary) intersects $\overline{AB}$ at X and $\overline{CD}$ at Y. We will show, in a sequence of steps, that $\overleftrightarrow{PF} \perp \overleftrightarrow{XY}$.

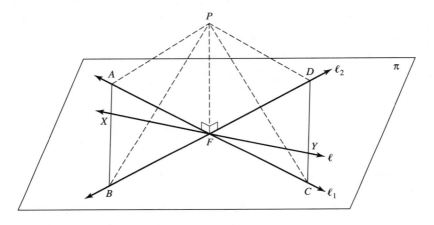

Figure 12.2: Line perpendicular to two lines in a plane

In the plane π (Fig. 12.3), the vertical angles $\angle AFB$ and $\angle DFC$ are congruent; thus $\triangle FAB \cong \triangle FCD$ by SAS. From corresponding parts, $\angle A \cong \angle C$

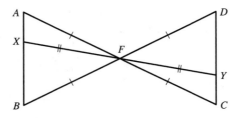

Figure 12.3: Triangles in π

and $\overline{AB} \cong \overline{CD}$. Putting the angle equivalence together with vertical angles at F, we find that $\triangle FAX \cong \triangle FCY$ by ASA, and again from corresponding parts, we have $FX = FY$ and $AX = CY$.

We were given that $\overleftrightarrow{PF}$ was perpendicular at F to each of ℓ_1 and ℓ_2, so $\angle PFA = \angle PFB = \angle PFC = \angle PFD = 90°$. Hence, we may use SAS to conclude that $\triangle PFA \cong \triangle PFB \cong \triangle PFC \cong \triangle PFD$ are four congruent right triangles and, moreover, that $PA = PB = PC = PD$. Applying this last relation together with $AB = CD$ from before, we have $\triangle PAB \cong \triangle PCD$ by SSS (Fig. 12.4). It follows that $\angle PAX \cong \angle PCY$ and thus by SAS that $\triangle PAX \cong \triangle PCY$. We are nearly done, since this last congruence yields $PX =$

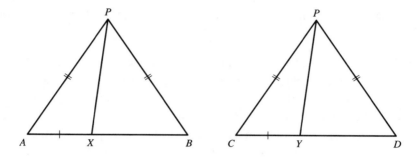

Figure 12.4: The slanted triangles

PY.

Now consider the $\triangle PXY$ with base on ℓ (Fig. 12.5). By SSS, $\triangle PXF \cong \triangle PYF$,

which means that $\angle PFX \cong \angle PFY$; however, $\angle XFY = \angle PFX + \angle PFY = 180°$ implies $\angle PFX = \angle PFY = 90°$. This shows that $\overline{PF} \perp \overline{XY}$ and the theorem is proved. $\qquad\square$

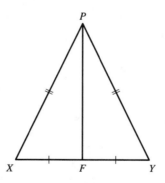

Figure 12.5: Triangle with base $\overline{XY}$ on ℓ

It will be useful to introduce language that allows us to paraphrase our last theorem. If ℓ is a line that intersects a plane π at a point F, then we will say ℓ is perpendicular to π at $F(\ell \perp \pi)$ if for any point Q on π we have $\ell \perp \overleftrightarrow{FQ}$. We may restate this theorem as follows: If $\overleftrightarrow{PF} \perp \overleftrightarrow{FA}$ and $\overleftrightarrow{PF} \perp \overleftrightarrow{FB}$, then $\overleftrightarrow{PF} \perp \pi(F, A, B)$.

We are now in a position to derive a number of corollaries from our work.

COROLLARY. (1) Let F be a point on a plane π and suppose points P_1 and P_2 are not on π. If $\overleftrightarrow{P_1F} \perp \pi$ and $\overleftrightarrow{P_2F} \perp \pi$, then $\overleftrightarrow{P_1F} = \overleftrightarrow{P_2F}$.

(2) Let F_1 and F_2 be points on a plane π and suppose point P is not on π. If $\overleftrightarrow{P_1F} \perp \pi$ and $\overleftrightarrow{P_2F} \perp \pi$, then $\overleftrightarrow{P_1F} = \overleftrightarrow{P_2F}$.

Notice that part (2) tells us that if we are able to find a line through P that is perpendicular to π, then it must be unique.

Proof. (1) We establish this part by contradiction. Suppose $\overleftrightarrow{P_1F} \neq \overleftrightarrow{P_2F}$. So these two distinct lines intersect at F and determine a unique plane $\pi' = \pi(\overleftrightarrow{P_1F}, \overleftrightarrow{P_2F})$. Now let ℓ be the line of intersection of π and π'. In particular, ℓ is in π so, by the previous theorem, $\overleftrightarrow{P_1F} \perp \ell$ at F and $\overleftrightarrow{P_2F} \perp \ell$ at F. But the three lines ℓ, $\overleftrightarrow{P_1F}$ and $\overleftrightarrow{P_2F}$ all lie in the plane π', which is impossible since in a plane there is only one perpendicular to a line at a given point. Thus, $\overleftrightarrow{P_1F} = \overleftrightarrow{P_2F}$.

(2) If F_1 and F_2 are distinct, then $\triangle PF_1F_2$ is a triangle with two right angles—an impossibility. $\qquad\square$

The next theorem marks the entry of metric notions into our study of solid geometry in this chapter.

THEOREM. Let $\overleftrightarrow{PF}$ be perpendicular to the plane π at F, P a point not on π, and suppose points A and B lie in π. Then $FA < FB$ if and only if $PA < PB$.

Proof. Apply the theorem of Pythagoras to the two triangles $\triangle PFA$ and $\triangle PFB$ and solve for PF^2 (Fig. 12.6).

We have $PF^2 = PA^2 - FA^2 = PB^2 - FB^2$, so rearrange terms to get $PB^2 - PA^2 = FB^2 - FA^2$. It is clear from this last equation that $PB - PA > 0$ if and only if $FB - FA > 0$. $\qquad\square$

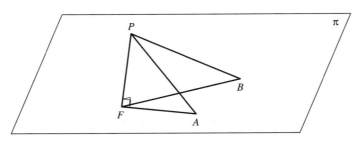

Figure 12.6: Distance from P to A and B

Observe under the conditions of the theorem that PF must be the shortest path length from P to π.

The next result offers a basic criterion for determining when two lines are parallel.

Theorem. Two distinct lines each perpendicular to the same plane are parallel.

Proof. Suppose ℓ_1 and ℓ_2 are perpendicular to the plane π at F_1 and F_2 respectively. It follows from a previous corollary that ℓ_1 and ℓ_2 do not intersect; thus, we must just show they lie in the same plane.

As in Fig. 12.7, choose a point A in π with $\overline{AF_1} \perp \overline{F_1F_2}$ and choose B on ℓ_2 such that $F_2B = F_1A$. Since $\ell_2 \perp \pi$, $\triangle BF_2F_1 \cong \triangle AF_1F_2$ by SAS, and

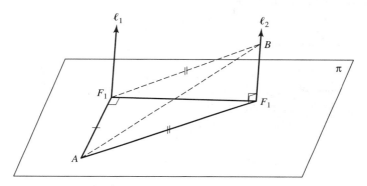

Figure 12.7: $\ell_1 \perp \pi$ and $\ell_2 \perp \pi$ imply $\ell_1 \| \ell_2$

therefore $F_1B = F_2A$. Now apply SSS to conclude that $\triangle F_1AB \cong \triangle F_2BA$ and, hence, that $\angle BF_2A = \angle BF_1A = 90°$. So, $\overleftrightarrow{AF_1} \perp \pi' = \pi(\overleftrightarrow{F_1F_2}, \overleftrightarrow{F_1B})$ at F_1, because $\overleftrightarrow{AF_1}$ is perpendicular to two lines that meet at F_1. By construction B and F_2 are in π'; we will be finished if we show that ℓ_1 is also in π'.

Assume ℓ_1 does not lie in π'. Set $\pi'' = \pi(F_2, \ell_1)$. Observe that $\overleftrightarrow{AF_1} \perp \ell_1$ and that $\overleftrightarrow{AF_1} \perp \overleftrightarrow{F_1F_2}$, which imply that $\overleftrightarrow{AF_1} \perp \pi''$ at F_1. Now it is clear that π intersects π' in $\overleftrightarrow{F_1F_2}$; moreover, suppose π intersects π'' in $\overleftrightarrow{F_1C}$, where C is on π'' but not on π'. However, it follows that both $\overline{F_1F_2}$ and $\overline{F_1C}$ are perpendicular at F_1 to $\overline{AF_1}$ in the plane π. This is a contradiction, for in a plane only one perpendicular may be constructed to a line at a point. Thus ℓ_1 must lie in π' and the proof is complete. $\qquad\square$

We can use the theorem to prove easily the two parts of the next corollary, which spell out relations concerning parallelism and perpendicularity.

Corollary. (1) If lines ℓ_1 and ℓ_2 are parallel and ℓ_1 is perpendicular to the plane π, then $\ell_2 \perp \pi$.

(2) If ℓ_1, ℓ_2, and ℓ_3 are lines such that $\ell_1 \| \ell_2$ and $\ell_2 \| \ell_3$, then $\ell_1 \| \ell_3$.

Proof. (1) Suppose ℓ_2 intersects π at P. If ℓ_2 is not perpendicular to π at P, then we can construct a distinct line ℓ_3, which is perpendicular to π at P. By our last theorem, we must have $\ell_1 \| \ell_3$. But this is impossible since $\ell_2 \| \ell_1$ and there is only one line through P parallel to ℓ_1. Hence, $\ell_2 \perp \pi$.

(2) Let π be any plane with $\ell_2 \perp \pi$. Part (1) of this corollary implies that $\ell_1 \perp \pi$ and, again, that $\ell_3 \perp \pi$. Now the theorem yields $\ell_1 \| \ell_3$.   □

We are now finally able to state and prove a theorem foreshadowed much earlier in this section. We have established the uniqueness of a perpendicular from a point to the plane, but we have never shown that such a perpendicular exists in the first place.

THEOREM. If P is a point not on the plane π, then there is a unique line ℓ through P such that $\ell \perp \pi$.

Proof. Let ℓ_1 be any line in π. In $\pi(P, \ell_1)$ construct a perpendicular to $\overline{PA}$ from P to ℓ_1 that meets ℓ_1 at A. In the plane π, let ℓ_2 be the line through A which is perpendicular to ℓ_1. Now in the plane $\pi(P, \ell_2)$ let ℓ be the line through P which is perpendicular to ℓ_2, say at F (Fig. 12.8).

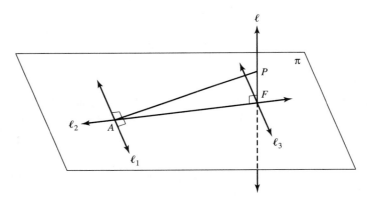

Figure 12.8: Perpendicular from P to π

In order to prove that $\ell \perp \pi$ we must find another line in π through F (in addition to ℓ_2) to which $\ell = \overleftrightarrow{PF}$ is perpendicular. To do this simply work in π to construct ℓ_3 through F parallel to ℓ_1. Since $\ell_1 \perp \ell_2$ and $\ell_1 \perp \overleftrightarrow{AP}$, we must have $\ell_1 \perp \pi(\overleftrightarrow{AP}, \overleftrightarrow{AF})$. Applying part (1) of the previous corollary, $\ell_1 \| \ell_3$ forces $\ell_3 \perp \pi(\overleftrightarrow{AP}, \overleftrightarrow{AF})$. But this means that $\overleftrightarrow{PF} \perp \ell_3$ and we conclude $\ell \perp \pi$, as desired. As we mentioned leading into the theorem, uniqueness of ℓ was already proved and we are finished.   □

As a consequence of this result we may define dist(P, π), the distance from a point P to a plane π. If P is not on π, then set dist$(P, \pi) = PF$, where $\overleftrightarrow{PF} \perp \pi$ at F; we have already observed that PF is the shortest distance from P to π. If P is on π, then define dist$(P, \pi) = 0$. We can squeeze a couple of other useful definitions in here as well. Define two planes π_1 and π_2 to be *parallel* (write $\pi_1 \| \pi_2$) if they do not intersect; finally, say that a line ℓ and a plane π are parallel (write $\ell \| \pi$) if they do not intersect.

The next lemma relates parallelism of planes to induced parallelism of certain lines and to perpendicularity.

LEMMA. Let π_1 and π_2 be parallel planes.
(1) If π is a plane that intersects π_1 in ℓ_1 and π_2 in ℓ_2, then $\ell_1 \| \ell_2$.
(2) If a line $\ell \perp \pi_1$, then $\ell \perp \pi_2$.

Proof. (1) Both ℓ_1 and ℓ_2 lie in π; however, if they intersect, then π_1 and π_2 also intersect, which is impossible. Thus, $\ell_1 \| \ell_2$.

(2) Suppose ℓ intersects π_1 at F_1 and intersects π_2 at F_2. Let A_1 be any other point of π_1 and set $\pi = \pi(A_1, \ell)$. It is clear that π intersects π_1 in $\overleftrightarrow{F_1 A_1}$ and suppose that π intersects π_2 in $\overleftrightarrow{F_2 A_2}$, where A_2 is a point of π_2. From part (1), $\overleftrightarrow{F_1 A_1} \| \overleftrightarrow{F_2 A_2}$ in π and, hence $\ell \perp \overleftrightarrow{F_2 A_2}$ because a line perpendicular to one of two parallel lines in a plane is perpendicular to the other. We may repeat this argument by choosing point B_1 of π_1 different from F_1 and A_1 and using it to find a point B_2 other than F_2 or A_2 of π_2 such that $\overleftrightarrow{F_1 B_1} \| \overleftrightarrow{F_2 B_2}$ and $\ell \perp \overleftrightarrow{F_2 B_2}$. Then $\ell \perp \pi(\overleftrightarrow{F_2 A_2}, \overleftrightarrow{F_2 B_2}) = \pi_2$. This completes the proof of the lemma. □

Our final theorem of this section allows us to make sense of the notion of distance between two parallel planes, for it tells us that parallel planes are everywhere equidistant.

THEOREM. If P_1 and Q_1 are points on the plane π_1, and P_2 and Q_2 are points on the plane π_2, and $\pi_1 \| \pi_2$, then $\mathrm{dist}(P_1, \pi_2) = \mathrm{dist}(Q_1, \pi_2) = \mathrm{dist}(P_2, \pi_1) = \mathrm{dist}(Q_2, \pi_1)$.

Proof. Denote the line perpendicular to π_1 at P_1 by ℓ_1 and the line perpendicular to π_2 at P_2 by ℓ_2. Suppose ℓ_1 intersects π_2 at F_2 and ℓ_2 intersects π_1 at F_1. From the lemma, $\ell_1 \perp \pi_2$ at F_2 which implies that $\ell_1 \| \ell_2$. In the plane of ℓ_1 and ℓ_2 (Fig. 12.9) we have $\overleftrightarrow{P_1 F_1} \| \overleftrightarrow{P_2 F_2}$, again from the lemma. Consequently, $P_1 F_2 P_2 F_1$ is a rectangle and the opposite sides $\overline{P_1 F_2}$ and $\overline{F_1 P_2}$ are congruent. Rephrasing this in the language of distance, we conclude that $\mathrm{dist}(P_1, \pi_2) = \mathrm{dist}(P_2, \pi_1)$. A similar approach will provide the proofs for the other two terms in the chain of equalities. □

Because of this last theorem, it now makes sense to define $\mathrm{dist}(\pi_1, \pi_2) = \mathrm{dist}(P, \pi_2)$ for any point P of π_1 in case the two planes π_1 and π_2 are parallel.

12.2   DIHEDRAL ANGLES

Let ℓ be a line in a plane π. In careful axiomatic treatments of Euclidean plane geometry either the following separation notion or something equivalent to it is postulated.

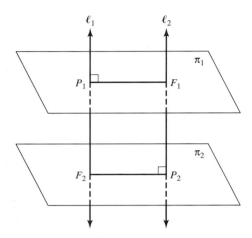

Figure 12.9: Distance between parallel planes

We will assume that the points of π other than ℓ are divided into two disjoint nonempty subsets that satisfy two properties:

(1) If P and Q are both in either one of these subsets, then the segment $\overline{PQ}$ is also in the subset.

(2) If P is in one of the subsets and Q in the other, then $\overline{PQ}$ intersects ℓ.

We call each of the subsets a half plane. Also, if P is a point of π in one of the subsets, then we refer to the half plane containing P as "the P side of ℓ "and (Fig. 12.10) denote it as $\pi_{\text{half}}(P, \ell)$.

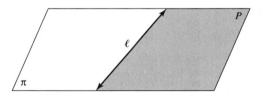

Figure 12.10: $\pi_{\text{half}}(P, \ell)$ (shaded) or the P side of ℓ

We will use the concepts of *parallel* and *antiparallel* rays. Suppose the two lines $\overleftrightarrow{AB}$ and $\overleftrightarrow{A'B'}$ are parallel. We will say that the ray $\overrightarrow{AB}$ is parallel to the ray $\overrightarrow{A'B'}$ if B and B' are in the same half plane determined by $\overleftrightarrow{AA'}$; say that $\overrightarrow{AB}$ and $\overrightarrow{A'B'}$ are antiparallel otherwise. Finally, we write $\overrightarrow{AB} \parallel \overrightarrow{A'B'}$ if these rays are parallel.

We will use the next lemma shortly.

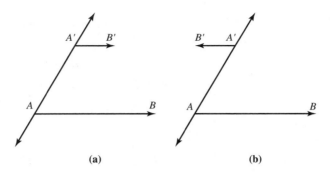

Figure 12.11: Parallel (a) and antiparallel (b) rays

LEMMA. If $\overrightarrow{AB} \parallel \overrightarrow{A'B'}$ and $\overrightarrow{AC} \parallel \overrightarrow{A'C'}$, then $\angle CAB = \angle C'A'B'$, where these angles are measured in $\pi(A, B, C)$ and $\pi(A', B', C')$, respectively.

Proof. We may assume that B, B', C, and C' are chosen such that $AB = A'B'$ and $AC = A'C'$. Consider the quadrilateral $AA'B'B$. Since the sides $\overline{AB}$ and $\overline{A'B'}$ are parallel and congruent, $AA'B'B$ must be a parallelogram. Hence, $BB' = AA'$ and $\overleftrightarrow{BB'} \parallel \overleftrightarrow{AA'}$. By the same argument $AA'C'C$ is also a parallelogram and so $\overleftrightarrow{CC'} \parallel \overleftrightarrow{AA'}$. Hence $\overline{BB'}$ and $\overline{CC'}$ are parallel and congruent to each other, so $BCC'B'$ is a parallelogram. Thus, $\overline{BC} \cong \overline{B'C'}$. Combining this fact with $\overline{AB} \cong \overline{A'B'}$, we see that $\triangle ABC \cong \triangle A'B'C'$ by SSS. Hence $\angle CAB = \angle C'A'B'$ as claimed.

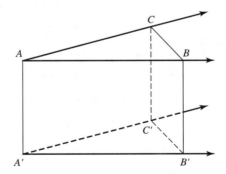

Figure 12.12: Angles at A and A' with parallel sides

□

A *dihedral angle* is a figure in space formed by two half planes $\pi_{\text{half}}(P, \ell)$ and $\pi_{\text{half}}(Q, \ell)$ that intersect in a common boundary line ℓ. We sometimes refer to each of the half planes in a dihedral angle as a *face* and to ℓ as the *edge*. Let A be a point on the edge ℓ of a dihedral angle. Choose B on one face so that $\overline{AB} \perp \ell$ and choose C

on the other face so that $\overline{AC} \perp \ell$. Note that $\pi(A, B, C)$ is perpendicular to ℓ. We call the angle $\angle BAC$ a *plane angle* of the dihedral angle.

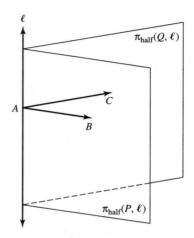

Figure 12.13: Plane angle $\angle BAC$ in a dihedral angle

LEMMA. All plane angles of a given dihedral angle are congruent.

Proof. Fix the notation for the dihedral angle as in the paragraph preceding this lemma; thus, $\angle BAC$ is one plane angle. Suppose A' is any other point on ℓ and B' and C' are chosen in $\pi_{\text{half}}(P, \ell)$ and $\pi_{\text{half}}(Q, \ell)$, respectively, as in Fig. 12.14 such that $\ell \perp \overleftrightarrow{A'B'}$ and $\ell \perp \overleftrightarrow{A'C'}$. We have $\ell \perp \pi(A, B, C)$ and $\ell \perp \pi(A', B', C')$, so $\pi(A, B, C) \| \pi(A', B', C')$. (See Exercise 12.8.) More-over, $\pi(P, \ell)$ is a cutting plane of both $\pi(A, B, C)$ and $\pi(A', B', C')$, which intersects them in $\overleftrightarrow{AB}$ and $\overleftrightarrow{A'B'}$ respectively. Consequently, $\overleftrightarrow{AB} \| \overleftrightarrow{A'B'}$, and since B and B' are on the P side of $\overleftrightarrow{AA'}$, $\overrightarrow{AB} \| \overrightarrow{A'B'}$. In a like manner, we can show that $\overrightarrow{AC} \| \overrightarrow{A'C'}$. It follows immediately from the preceding lemma that $\angle BAC = \angle B'A'C'$. $\qquad\square$

This last lemma justifies defining the size or measure of a dihedral angle to be the measure of any one of its plane angles. Two dihedral angles are said to be congruent if they have equal sizes. The proof of the following theorem is now very easy.

THEOREM. If two plane angles intersect, then the vertical dihedral angles formed are congruent.

Proof. Just note that any cutting plane perpendicular to the line of intersection contains two pairs of equal vertical angles. $\qquad\square$

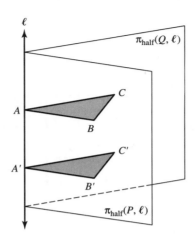

Figure 12.14: Two plane angles in a dihedral angle

It follows almost immediately from the theorem that if two planes intersect and any one of the dihedral angles formed measures 90°, then all four dihedral angles measure 90°. In this case we say that the two planes are *perpendicular*.

THEOREM. If π_1 and π_2 are planes and ℓ is a line contained in π_2 with $\ell \perp \pi_1$, then $\pi_2 \perp \pi_1$.

Proof. Let $\ell \perp \pi_1$ at F and suppose ℓ' is the line of intersection of π_1 and π_2 (Fig. 12.15). Choose a point P on π_1 on either side of ℓ' so that $\overleftrightarrow{FP} \perp \ell'$ at F and choose any point G on ℓ. From $\overleftrightarrow{FG} = \ell \perp \pi_1$, we have $\ell' \perp \overleftrightarrow{FG}$ and thus $\ell' \perp \pi(F, G, P)$. Therefore, $\angle PFG$ is a plane angle of the dihedral angle with edge ℓ' and faces $\pi_{\text{half}}(P, \ell')$ and $\pi_{\text{half}}(G, \ell')$. Finally $\angle PFG = 90°$, which implies one dihedral angle, and therefore all four dihedral angles, are right angles. Thus $\pi_1 \perp \pi_2$, as claimed.   □

12.3 PROJECTIONS

If π is a plane and P is a point not on π, then we know there is a unique line through P which is perpendicular to π. Denote the point F, at which the perpendicular intersects π, by $\text{proj}_\pi(P)$ and call F the *projection* of P on π. If P is already on π, then we set $\text{proj}_\pi(P) = P$. We can extend our projection operator, which is a function mapping points to points, to a related operator which maps lines (which are certain sets of points) to lines. To do this for any line ℓ, we define $\text{proj}_\pi(\ell)$ to consist of the projections on π of all points of ℓ. In set notation, the projection of ℓ on $\pi = \text{proj}_\pi(\ell) = \{\text{proj}_\pi(P) | P$ is a point on $\ell\}$. From what has been said in this paragraph, it quickly follows that if $\ell \perp \pi$, then $\text{proj}_\pi(\ell)$ is the single point at which ℓ intersects π. (There is a pedantic—and correct—distinction to be made between a

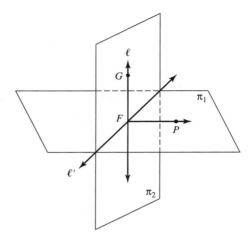

Figure 12.15: Plane perpendicularity criterion

point and the set whose sole element is that point. The distinction is not in the spirit of
this text and will not concern us here.)

THEOREM. If π is a plane and ℓ a line not perpendicular to π, then $\text{proj}_\pi(\ell)$ is
a line contained in π.

Proof. For P a point on ℓ but not on π, let $P' = \text{proj}_\pi(P)$. Then $\overleftrightarrow{PP'} \perp \pi$
and, by the final theorem of Section 12.2, $\pi' = \pi(\overleftrightarrow{PP'}, \ell) \perp \pi$. (Fig. 12.16)
Suppose ℓ' is the line of intersection of π' and π. We claim that $\text{proj}_\pi(\ell) = \ell'$.
To see this, first assume Q is any other point on ℓ and that Q' is the foot of the
perpendicular from Q to ℓ' in the plane π'. Choose any point R on π so that
$\overleftrightarrow{RQ'} \perp \ell'$ at Q'. It must be the case that $\angle QQ'R = 90°$, for this angle is a

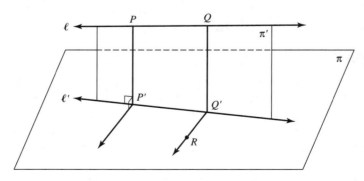

Figure 12.16: $\ell' = Proj_\pi(\ell)$

plane angle of one of the dihedral angles formed by the perpendicular planes π'

and π. So $\overleftrightarrow{QQ'}$ is perpendicular to both $\overleftrightarrow{Q'R}$ and ℓ' and thus $\overleftrightarrow{QQ'} \perp \pi$ at Q'. Therefore, $Q' = \text{proj}_\pi(Q)$ and we have shown that any point of ℓ projects into ℓ'.

We will be finished if we show that any point in ℓ' is the projection of some point of ℓ. Assume Q' is an arbitrary point of ℓ'. At Q', erect the unique line perpendicular to π. This perpendicular to π at Q' must lie in π' since lines perpendicular to the same plane are parallel; therefore, it must intersect ℓ (otherwise $\ell \perp \pi$) at a unique point Q. By this construction, $\text{proj}_\pi(Q) = Q'$ and we are finished with the claim and the proof. $\square$

The next result follows from the proof of the theorem.

COROLLARY. If π is a plane and ℓ a line not perpendicular to π, then there is a unique plane π' containing ℓ such that $\pi' \perp \pi$.

We can now derive another nice relationship between lines and their projections upon planes. However, in order to prove the theorem, we need a plane geometry result that is a "sublemma" for us. We leave the proof of this result as a challenge for your. Here it is: If $\triangle ABC$ and $\triangle A'B'C'$ have right angles at C and C', respectively, $AB = A'B'$ and $BC < B'C'$, then $\angle BAC < \angle B'A'C'$. (The theorem of Pythagoras might be helpful here.) We now proceed to the theorem mentioned.

THEOREM. Let ℓ be a line which intersects the plane π at A with ℓ not perpendicular to π. Then the acute angle between ℓ and $\text{proj}_\pi(\ell)$ is less than the angle between ℓ and any other line of π passing through A.

Proof. Let P be any other point of ℓ and set $P' = \text{proj}_\pi(P)$ as in Fig. 12.17. Observe that P' is on $\ell' = \text{proj}_\pi(\ell)$. Now let ℓ'' be any line of π through A that is distinct from ℓ'. If Q is on ℓ'', then we must prove that $\angle PAP' < \angle PAQ$. If $PAQ \geq 90°$, then this is clear since $\angle PAP' < 90°$. So we might as well

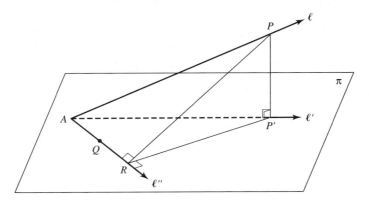

Figure 12.17: $\angle PAR > \angle PAP'$

assume that $\angle PAQ$ is acute also.

Suppose the line through P that is perpendicular to ℓ'' intersects ℓ'' at R. From $\overleftrightarrow{PP'} \perp \pi$ we have $\angle PP'R = 90°$ and therefore $PR > PP'$. Now apply our sublemma to the right triangles $\triangle APP'$ and $\triangle APR$ since they have the hypotenuse $\overline{AP}$ in common. We conclude that $\angle PAP' < \angle PAQ$, which proves the theorem.   □

Our last theorem in this section is a striking result that exhibits a distance property of skew lines.

THEOREM. If ℓ and ℓ'' are skew lines, then there are points P on ℓ and P' on ℓ'' that have the following three properties:

(1) $\overleftrightarrow{PP'} \perp \ell$ and $\overleftrightarrow{PP'} \perp \ell''$.

(2) $\overleftrightarrow{PP'}$ is the unique line satisfying (1).

(3) $\overline{PP'}$ is the shortest segment from a point on ℓ to a point on ℓ''.

Proof. Initially we assert that there is a unique plane π containing ℓ'' such that $\ell \| \pi$. To see this, pick any point T of ℓ'' and construct the unique line ℓ_1 through T such that $\ell_1 \| \ell$. Set $\pi = \pi(\ell_1, \ell'')$. If ℓ punctured π it would have to do so in the unique plane π_1 determined by ℓ and ℓ_1 and would thus intersect ℓ_1, which is the line of intersection of π_1 and π. (Also see Exercise 4 of this chapter.)

Denote $\text{proj}_\pi(\ell)$ by ℓ' and note that ℓ' cannot be parallel to ℓ''; for otherwise, $\ell' \| \ell''$ and $\ell' \| \ell$ (by the proof of the previous theorem) would imply that $\ell \| \ell''$, a contradiction. Suppose ℓ' intersects ℓ'' at the unique point P' on π. Let P be the unique point on ℓ with $\text{proj}_\pi(P) = P'$. Consequently $\overleftrightarrow{PP'} \perp \ell'$ and $\overleftrightarrow{PP'} \perp \ell$ which establishes property (1) of the theorem (Fig. 12.18.).

To derive (2) we assume the contrary, namely for some points Q on ℓ and R on ℓ'' we have $\overleftrightarrow{QR} \perp \ell$ and $\overleftrightarrow{QR} \perp \ell''$. Through the point R construct the line $\overleftrightarrow{RS}$ in π such that $\overleftrightarrow{RS} \| \ell'$. Because $\ell' \| \ell$, $\overleftrightarrow{QR} \perp \overleftrightarrow{RS}$. Thus $\overleftrightarrow{QR} \perp \pi(\overleftrightarrow{RS}, \ell'') = \pi$. However, $\overleftrightarrow{QQ'} \perp \pi$, where $Q' = \text{proj}_\pi(Q)$; but there is only one perpendicular from Q to π, so we have a contradiction. Hence $\overleftrightarrow{PP'}$ is the unique line satisfying (1), and (2) is proved.

In order to prove part (3) of the theorem we again make use of the picture in Fig. 12.18 with a different meaning for the segment $\overleftrightarrow{QR}$. Here, let $\overleftrightarrow{QR}$ be any segment from a point Q on ℓ to a point R on ℓ''. We want to show that $PP' < QR$. With $Q' = \text{proj}_\pi(Q)$, it follows from $\ell \| \ell''$ that $PP' = QQ'$. In the right $\triangle RQQ'$ the hypotenuse $QR > QQ'$, which proves (3).   □

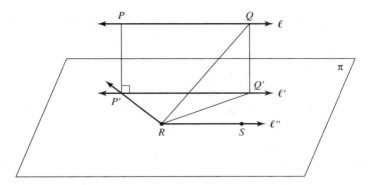

Figure 12.18: $PP' =$ shortest distance between ℓ and ℓ''

12.4   TRIHEDRAL ANGLES

The presentation of trihedral angles in this section is made both for their intrinsic interest and for use with spherical triangles in the next chapter.

A *trihedral angle* $\angle A B_1 B_2 B_3$ is formed when the edges of three half planes meet in a common point called the vertex and consists of this vertex, A; edges $\overrightarrow{AB_1}$, $\overrightarrow{AB_2}$ and $\overrightarrow{AB_3}$; and faces composed of points of the half planes $\pi_{\text{half}}(\overleftrightarrow{AB_1}, B_2)$, $\pi_{\text{half}}(\overleftrightarrow{AB_2}, B_3)$, and $\pi_{\text{half}}(\overrightarrow{AB_3}, B_1)$ bounded by the edges. (Fig. 12.19(a)). We call the three angles $\angle B_1 A B_2$, $\angle B_2 A B_3$, and $\angle B_3 A B_1$ face angles. Note that the vertex comes first in the notation for the trihedral angle and then points on consecutive edges. Observe further

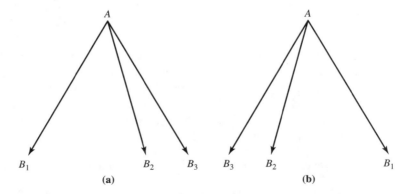

Figure 12.19: Symmetric trihedral angles $\angle A B_1 B_2 B_3$ and $\angle A B_3 B_2 B_1$

that a trihedral angle is composed of portions of three dihedral angles; in Fig. 12.19(a), the ones between $\pi_{\text{half}}(\overrightarrow{AB_1}, B_2)$ and $\pi_{\text{half}}(\overrightarrow{AB_1}, B_3)$, between $\pi_{\text{half}}(\overrightarrow{AB_2}, B_1)$ and $\pi_{\text{half}}(\overrightarrow{AB_2}, B_3)$, and between $\pi_{\text{half}}(\overrightarrow{AB_3}, B_2)$ and $\pi_{\text{half}}(\overrightarrow{AB_3}, B_1)$.

An inequality lemma from plane geometry, sometimes called the scissors lemma, will come in handy for our first trihedral result and we start with it. Its proof is not easy.

> **LEMMA.** In $\triangle ABC$ and $\triangle A'B'C'$, if $\overline{AB} \cong \overline{A'B'}$ and $\overline{AC} \cong \overline{A'C'}$, then $\angle A > \angle A'$ if and only if $BC > B'C'$.

Proof. Since this is an "if and only if" result we must provide proofs for two directions. We first assume that $\angle A > \angle A'$ and show $BC > B'C'$; following this we do the converse.

First we build a copy of $\triangle A'B'C'$ in $\triangle ABC$ as in Fig. 12.20. Because $\angle A > \angle A'$ we can find a point D in the interior of $\angle ABC$ such that $\angle DAC = \angle B'A'C'$ and $DA = BA = B'A'$. By SAS, $\triangle A'B'C' \cong \triangle ADC$ and $DC = B'C'$. Now let E be the point on $\overline{BC}$ such that $\overline{AE}$ is the bisector of $\angle BAD$. From SAS

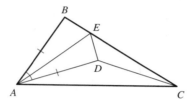

Figure 12.20: Scissors inequality

again, we have $\triangle ABE \cong \triangle ADE$, from which it follows that $BE = ED$.

From the triangle inequality applied to $\triangle DEC$ we have that $ED + EC > DC$. Hence, $BE + EC > DC$, or $BC > DC = B'C'$, which we wanted to prove.

To establish the converse we assume $CB > C'B'$ and show that $\angle A > \angle A'$. Observe that $\angle A \neq \angle A'$, for otherwise $\triangle ABC \cong \triangle A'B'C'$ by SAS and we would have $BC = B'C'$, a contradiction. We now proceed by assuming the contrary and deriving a contradiction. Assume that $\angle A' > \angle A$. Now simply apply the half of the lemma we have already proved to conclude that $B'C' > BC$, a contradiction. Thus, $\angle A > \angle A'$ and the lemma is proved.   □

In fact, we only use one direction of this lemma in the first theorem of this section.

> **THEOREM.** The sum of any two face angles of a trihedral angle is greater than the third face angle.

Proof. There is at least one face anglethat is greater than or equal to each of the other two. Arrange the lettering so that this "largest" face angle is $\angle B_3 A B_1$ of the trihedral angle $\angle A B_1 B_2 B_3$ as in Fig. 12.21. In order to prove the theorem it suffices to show that $\angle B_1 A B_2 + \angle B_2 A B_3 > \angle B_3 A B_1$. (Why?) Since $\angle B_3 A B_1 \leq \angle B_1 A B_2$ there is a point C in $\pi(A, B_1, B_3)$ and in the face of the trihedral angle

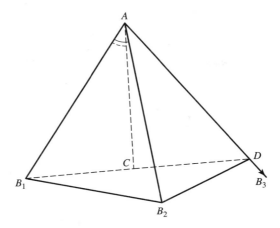

Figure 12.21: Face angles in a trihedral angle

such that $\angle B_1 AC = \angle B_1 AB_2$ and $AC = AB_2$. By SAS, $\triangle B_1 AB_2 \cong \triangle B_1 AC$, and therefore $B_1 B_2 = B_1 C$.

Extend $\overline{B_1 C}$ to intersect $\overline{A B_3}$ in D and use the triangle inequality in $\triangle B_1 B_2 D$ to conclude that $B_1 B_2 + B_2 D > B_1 D$, or $B_1 B_2 + B_2 D > B_1 C + C D$. Since $B_1 B_2 = B_1 C$ we may subtract to obtain $B_2 D > C D$. Now we can apply the previous lemma to the triangles $\triangle B_2 AD$ and $\triangle CAD$. We have $AB_2 = AC$, $\overline{AD}$ a common side, and $B_2 D > DC$, so the lemma implies that $\angle B_2 AD > \angle CAD$. By addition, $\angle B_1 AB_2 + \angle B_2 AD > \angle B_1 AB_2 + \angle CAD$ or, after substitution, $\angle B_1 AB_2 + \angle B_2 AB_3 > \angle B_1 AC + \angle CAD = \angle B_1 AD = \angle B_1 AB_3$. This proves the theorem. $\qquad \square$

The next theorem is in the same spirit as our last one and carries it one step further.

THEOREM. The sum of the face angles of any trihedral angle is less than $360°$.

Proof. Let the trihedral angle be $\angle AB_1 B_2 B_3$ and consider the resulting solid figure bounded by its three faces and by the $\triangle B_1 B_2 B_3$ (Fig. 12.22). This figure is called a *tetrahedron* (four faces), and note that it has four trihedral angles. We apply the previous theorem to the three trihedral angles with vertices at B_1, B_2, and B_3 to yield the following three inequalities:

$$\angle AB_2 B_1 + \angle AB_2 B_3 > \angle B_1 B_2 B_3$$

$$\angle AB_3 B_2 + \angle AB_3 B_1 > \angle B_2 B_3 B_1$$

$$\angle AB_1 B_3 + \angle AB_1 B_2 > \angle B_3 B_1 B_2.$$

We now add these inequalities, use the fact that the resulting right-hand side is the angle sum of $\triangle B_1 B_2 B_3$, and rearrange the terms on the left-hand side to see

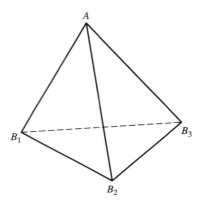

Figure 12.22: A tetrahedron

that

$$(\angle AB_2B_1 + \angle AB_1B_2) + (\angle AB_2B_3 + \angle AB_3B_2) + (\angle AB_3B_1 + \angle AB_1B_3) > 180°.$$

The sum in the first parentheses is $180° - \angle B_1AB_2$, in the second is $180° - \angle B_2AB_3$, and in the third is is $180° - \angle B_3AB_1$; substitute and transpose terms to see that

$$\angle B_1AB_2 + \angle B_2AB_3 + \angle B_3AB_1 < 360°,$$

which we wanted to prove. □

We will finish this section with two results about congruence of trihedral angles. We say that two trihedral angles $\angle AB_1B_2B_3$ and $\angle A'B_1'B_2'B_3'$ are congruent if the following six conditions hold (for $B_1'B_2'B_3'$ taken in the same consecutive ordering as $B_1B_2B_3$):

1. $\angle B_1AB_2 \cong \angle B_1'A'B_2'$

2. $\angle B_2AB_3 \cong \angle B_2'A'B_3'$

3. $\angle B_3AB_1 \cong \angle B_3'A'B_1'$

4. the dihedral angle with edge $\overrightarrow{AB_1} \cong$ the dihedral angle with edge $\overrightarrow{A'B_1'}$

5. the dihedral angle with edge $\overrightarrow{AB_2} \cong$ the dihedral angle with edge $\overrightarrow{A'B_2'}$

6. the dihedral angle with edge $\overrightarrow{AB_3} \cong$ the dihedral angle with edge $\overrightarrow{A'B_3'}$

The essential idea here is that trihedral angles are congruent if face angles and dihedral angles of one are equal to corresponding face angles and dihedral angles of the other when arranged in the same order. On the other hand, if face angles and dihedral angles of one trihedral angle are equal to face angles and dihedral angles of another arranged in opposite order, then the two trihedral angles are mirror images of each other (Fig. 12.19 (a) and (b)) and we say they are *symmetric*. You are strongly advised to get out your scissors, tape, and paper to construct models in order to be convinced that congruent trihedral angles can be made to coincide whereas symmetric ones cannot.

THEOREM. If the three face angles of one trihedral angle are congruent respectively to the three face angles of another trihedral angle, then the two trihedral angles are either congruent or symmetric.

Proof. We are given trihedral angles with $\angle B_1 A B_2 = \angle B_1' A' B_2'$, $\angle B_2 A B_3 = \angle B_2' A' B_3'$, and $\angle B_3 A B_1 = B_3' A' B_1'$. We lose no generality by making the assumption that $A B_1 = A B_2 = A B_3 = A' B_1' = A' B_2' = A' B_3'$. By SAS we have the three congruences $\triangle A B_1 B_2 \cong \triangle A' B_1' B_2'$, $\triangle A B_2 B_3 \cong \triangle A' B_2' B_3'$, and $\triangle A B_3 B_1 \cong \triangle A' B_3' B_1'$. Note that the congruences here and to follow are from either (a) to (b) or from (a) to (c) in Fig. 12.23. Thus, $B_1 B_2 = B_1' B_2'$, $B_2 B_3 = B_2' B_3'$, and $B_3 B_1 = B_3' B_1'$; which imply $\triangle B_1 B_2 B_3 \cong \triangle B_1' B_2' B_3'$ from SSS. From any point C of $\overline{A B_1}$ construct $\overline{C D}$ in $\pi(A, B_1, B_2)$ with D on $\overline{B_1 B_2}$ such that $\overline{C D} \perp \overline{A B_1}$. Similarly, construct $\overline{C E}$ in $\pi(A, B_1, B_3)$ with E on $\overline{B_1 B_3}$ with $\overline{C E} \perp \overline{A B_1}$. The angle $\angle D C E$ is a plane angle of the dihedral angle with edge $\overrightarrow{A B_1}$. Let C' be the point on $\overrightarrow{A' B_1'}$ such that $C B_1 = C' B_1'$ and let $\angle D' C' E'$ be the plane angle of the dihedral angle with edge $\overrightarrow{A' B_1'}$ with vertex C', and with D' on $\overline{B_1' B_2'}$ and E' on $\overline{B_1' B_3'}$.

From corresponding parts we have $\angle A B_1 B_2 = \angle A' B_1' B_2'$ and $\angle A B_1 B_3 = \angle A' B_1' B_3'$ and, from ASA, we have the right triangle congruences $\triangle B_1 C D \cong \triangle B_1' C' D'$ and $\triangle B_1 C E \cong \triangle B_1' C' E'$. Hence, $B_1 D = B_1' D'$ and $B_1 E = B_1' E'$ as corresponding parts of these latter triangles. Again, from the previously observed congruence of the base triangles $\triangle B_1 B_2 B_3$ and $\triangle B_1' B_2' B_3'$, we know $\angle D B_1 E = \angle D' B_1' E'$. Consequently, SAS yields $\triangle B_1 D E \cong \triangle B_1' D' E'$ and, therefore, $D E = D' E'$. Put this last equality together with $C D = C' D'$ and $C E = C' E'$, which arise from our former right triangle congruences, to conclude that $\triangle C D E \cong \triangle C' D' E'$, by SSS. Finally, we may conclude that $\angle D C E = \angle D' C' E'$ and, since these are plane angles of the dihedral angles with edges at $\overrightarrow{A B_1}$ and $\overrightarrow{A' B_1'}$, that these dihedral angles are equal.

In a similar manner it may be shown that the dihedral angles with edges at $\overrightarrow{A B_2}$ and $\overrightarrow{A' B_2'}$, and at $\overrightarrow{A B_3}$ and $\overrightarrow{A' B_3'}$ are equal. Depending upon whether the consecutive orderings of these face and dihedral angles are from (a) to (b) or from (a) to (c) of Figure 12.23, we have now proved the trihedral angles to be either congruent or symmetric.   □

We conclude with one more theorem in the same spirit.

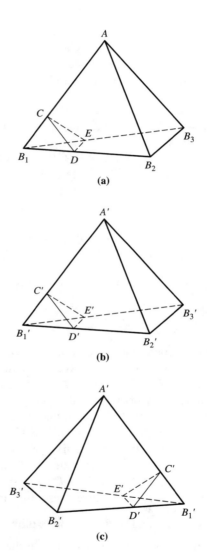

Figure 12.23: Trihedral in (a) congruent to (b), symmetric to (c)

THEOREM. Given trihedral angles $\angle AB_1B_2B_3$ and $\angle A'B_1'B_2'B_3'$ such that the dihedral angle at $\overrightarrow{AB_1}$ is congruent to the dihedral angle at $\overrightarrow{A'B_1'}$, $\angle B_1AB_2 \cong \angle B_1'A'B_2'$ and $\angle B_1AB_3 \cong \angle B_1'A'B_3'$. Then the two trihedral angles are either congruent or symmetric.

Proof. In order to prove that the two trihedral angles are congruent, we will prove that $\angle B_2AB_3 \cong \angle B_2'A'B_3'$. The result will then follow from the previous theorem.

First, we may choose B_1, B_1' so that $\overline{AB_1} \cong \overline{A'B_1'}$. Also we may choose B_2, B_3, B_2', B_3' so that $\overline{B_2B_1}$ and $\overline{B_3B_1}$ will be perpendicular to $\overline{AB_1}$; and $\overline{B_2'B_1'}$ and $\overline{B_3'B_1'}$ will be perpendicular to $\overrightarrow{A'B_1'}$. With these choices $\triangle AB_1B_2 \cong \triangle A'B_1'B_2'$ and $\triangle AB_1B_3 \cong \triangle A'B_1'B_3'$, by ASA in both cases. Hence $\overline{B_1B_2} \cong \overline{B_1'B_2'}$ and $\overline{B_1B_3} \cong \overline{B_1'B_3'}$.

Next, by the definition of congruent dihedral angles, it follows that $\angle B_2B_1B_3 \cong \angle B_2'B_1'B_3'$. Hence, $\triangle B_2B_1B_3 \cong \triangle B_2'B_1'B_3'$, by SAS.

Finally, consider the triangles $\triangle AB_2B_3$ and $\triangle A'B_2'B_3'$. The sides $\overline{B_2B_3}$ and $\overline{B_2'B_3'}$ are congruent because $\triangle B_1B_2B_3 \cong \triangle B_1'B_2'B_3'$; the sides $\overline{AB_2}$ and $\overline{A'B_2'}$ are congruent because $\triangle AB_1B_3 \cong \triangle A'B_1'B_3'$. So $\triangle AB_2B_3 \cong \triangle A'B_2'B_3'$, thus $\angle B_2AB_3 \cong \angle B_2'A'B_3'$ and we are done. $\qquad\square$

PROBLEMS

12.1. Show that if P is a point that is not on the plane π with $\overleftrightarrow{PF} \perp \pi$ at F and there is a line $\ell \perp \overleftrightarrow{PF}$ at F, then ℓ lies in π.

12.2. If $\overleftrightarrow{AB}$ is a line in space and π is the set of points in space that includes A as well as all points C such that $\angle BAC$ is a right angle (i.e., $\pi = \{A\} \cup \{C \mid \angle BAC = 90°\}$), then prove that π is the plane through A that is perpendicular to $\overleftrightarrow{AB}$.

12.3. If ℓ is a line and P a point not on ℓ, then prove that there is a unique plane π containing P such that $\ell \perp \pi$.

12.4. Prove that if ℓ_1 and ℓ_2 are parallel lines and ℓ_1 is in plane π, then either ℓ_2 is in π or $\ell_2 \| \pi$.

12.5. Suppose that line ℓ_1 intersects line ℓ_2 and π is a plane such that $\ell_1 \| \pi$ and $\ell_2 \| \pi$. Show that $\pi(\ell_1, \ell_2) \| \pi$.

*12.6. Prove that if two lines are parallel, then there is a plane containing the first line and parallel to the second.

*12.7. Let ℓ_1 and ℓ_2 be skew lines and let P be a point on neither ℓ_1 nor ℓ_2. Construct the unique plane containing P which is parallel to each of ℓ_1 and ℓ_2.

12.8. Prove that two distinct planes perpendicular to the same line are parallel.

*12.9. If π_1, π_2, and π_3 are planes with $\pi_1 \| \pi_2$ and $\pi_2 \| \pi_3$, then prove that $\pi_1 \| \pi_3$.

12.10. Provide a proof for the following theorem. If two parallel planes are cut by a transversal plane, then the alternate interior dihedral angles formed are congruent. Conversely, if two planes are intersected by a transversal plane such that the alternate interior dihedral angles are the same size, then the two planes are parallel.

12.11. Show that a plane perpendicular to each of two intersecting planes is itself perpendicular to their line of intersection.

12.12. Prove that the projection of a parallelogram to a plane is either a parallelogram or a line segment.

12.13. Show that the ratio of the lengths of the projections to a plane of two parallel line segments is equal to the ratio of the lengths of these segments.

*12.14. Let A, B, C and D be any four points in space. (You can think of them as the vertices of a tetrahedron.) Let E be the midpoint of $\overline{AB}$, F the midpoint of $\overline{BC}$, G the midpoint of $\overline{CD}$, and H the midpoint of $\overline{DA}$. Prove that $EFGH$ is a parallelagram.

12.15. Prove that two trihedral angles are equal if the three dihedral angles of each are right dihedral angles.

12.16. Show that in a trihedral angle, the three planes bisecting the three face angles perpendicular to their respective planes intersect in a line, every point of which is equidistant from the three edges of the trihedral angle.

12.17. Let ℓ be a line that intersects the plane π at A, ℓ not perpendicular to π. Then the obtuse angle between ℓ and $\text{proj}_\pi(\ell)$ is greater than the angle between ℓ and any other line of π passing through A.

*12.18. Let ℓ and ℓ' be skew lines and A a point not on ℓ or ℓ'. Prove that there is a unique line through A which intersects both ℓ and ℓ'.

12.19. If $\angle AB_1 B_2 B_3 B_4$ is a quadrahedral angle, prove that $\angle B_1 A B_2 + \angle B_2 A B_3 + \angle B_3 A B_4 + \angle B_4 A B_1 < 360°$.

12.20. A solid figure with four vertices is called a tetrahedron (plural, tetrahedra). A tetrahedron $ABCD$ has four faces, $\triangle ABC$, $\triangle ACD$, $\triangle ADB$, and $\triangle BCD$, which together have six edges and twelve face angles. It also has dihedral angles, one at each edge. Two tetrahera $ABCD$ and $A'B'C'D'$ are congruent if all 24 corresponding parts are congruent. Prove that $ABCD \cong A'B'C'D'$ in each of these cases:

   (a)   All corresponding edges are congruent.

   (b)   All corresponding face angles and one pair of corresponding edges are congruent.

   (c)   Trihedral angles $\angle ABCD \cong \angle A'B'C'D'$ and $\overline{AB} \cong \overline{A'B'}$, $\overline{AC} \cong \overline{A'C'}$ and $\overline{AD} \cong \overline{A'D'}$.

12.21. **(a)**   Continuing with the ideas of Exercise 19, write a definition of when two tetrahedra are similar.

   (b)   Use your answer to (a) to prove that $ABCD \sim AB'C'D'$ if the planes $\pi(B, C, D)$ and $\pi(B'C'D')$ are parallel.

CHAPTER SUMMARY

- We started out this chapter with three axioms.

  (S1) There is a unique plane containing three given noncollinear points. If A, B, and C are the given points, then we will denote the plane they determine by $\pi(A, B, C)$.

  (S2) If two distinct points are in a plane, then the entire line they determine also is in the plane.

  (S3) If two planes intersect, then their intersection consists of more than one point.

- As immediate consequences of the axioms, we added the following.

  1. If ℓ is a line and P is a point not on ℓ, then there is a unique plane π which contains P and ℓ, often denoted $\pi(P, \ell)$.

  2. If ℓ_1 and ℓ_2 are distinct intersecting lines, then there is a unique plane π that contains both ℓ_1 and ℓ_2, often denoted $\pi(\ell_1, \ell_2)$.

  3. There is at most one plane containing two distinct lines.

- Lines that do not intersect and are not contained in the same plane are skew lines. Lines that do not intersect and are contained in the same plane are called parallel lines.

  Next, we had the following theorems and corollaries.

- If two distinct planes intersect, then their intersection is a line.

- Let ℓ_1 and ℓ_2 be distinct lines in the plane π that intersect at the point F. If P is a point that is not on π with $\overleftrightarrow{PF} \perp \ell_1$ and $\overleftrightarrow{PF} \perp \ell_2$, then $\overleftrightarrow{PF}$ is perpendicular to any line in π that passes through F, as shown in the following figure. In such an event, we say that $\overleftrightarrow{PF}$ is perpendicular to π.

- Let F be a point on a plane π and suppose points P_1 and P_2 are not on π. If $\overleftrightarrow{P_1F} \perp \pi$ and $\overleftrightarrow{P_2F} \perp \pi$, then $\overleftrightarrow{P_1F} = \overleftrightarrow{P_2F}$.

- Let F_1 and F_2 be points on a plane π and suppose point P is not on π. If $\overleftrightarrow{PF_1} \perp \pi$ and $\overleftrightarrow{PF_2} \perp \pi$, then $\overleftrightarrow{PF_1} = \overleftrightarrow{PF_2}$.

- Let $\overleftrightarrow{PF}$ be perpendicular to the plane π at F, P a point not on π, and suppose points A and B lie in π. Then $FA < FB$ if and only if $PA < PB$, as depicted in the figure on page 172.

- If P is a point that is not on the plane π with $\overleftrightarrow{PF} \perp \pi$ at F and there is a line $\ell \perp \overleftrightarrow{PF}$ at F, then ℓ lies in π.

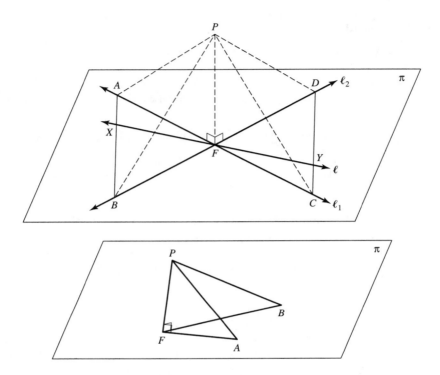

- If $\overleftrightarrow{AB}$ is a line in space and π is the set of all points in space that includes A and all points C such that $\angle BAC$ is a right angle, then π is the plane through A which is perpendicular to $\overleftrightarrow{AB}$.

- Two distinct lines are parallel when they are perpendicular to the same plane.

- If lines ℓ_1 and ℓ_2 are parallel and ℓ_1 is perpendicular to the plane π, then $\ell_2 \perp \pi$.

- If ℓ_1, ℓ_2, and ℓ_3 are lines such that $\ell_1 \| \ell_2$ and $\ell_2 \| \ell_3$, then $\ell_1 \| \ell_3$.

- If P is a point not on the plane π, then there is a unique line ℓ through P such that $\ell \perp \pi$.

- Define the distance from P to π as the length of the segment $\overline{PF}$, where F is the foot of this perpendicular; dist(P, π) is the notation for the distance from a point P to a plane π.

- Define $\pi_1 \| \pi_2$ if π_1 and π_2 do not intersect.

  A line ℓ and a plane π are parallel if they do not intersect.

- Let π_1 and π_2 be parallel planes. If π is a plane which intersects π_1 in ℓ_1 and π_2 in ℓ_2, then $\ell_1 \| \ell_2$. If a line $\ell \perp \pi_1$, then $\ell \perp \pi_2$.

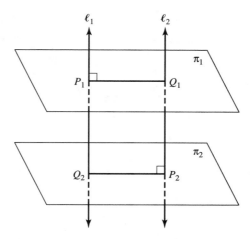

If P_1 and Q_1 are points on the plane π_1, P_2 and Q_2 are points on the plane π_2, and $\pi_1 \| \pi_2$, then dist(P_1, π_2) =dist(Q_1, π_2) =dist(P_2, π_1) =dist(Q_2, π_1), as shown in the preceding figure.

- Let ℓ be a line in a plane π. We take as an axiom that the points of π other than ℓ are divided into nonempty subsets that satisfy the following properties:

    1. If P and Q are both in either one of the subsets, then the segment $\overline{PQ}$ is also in the subset [as pictured in (a) in the following figure].
    2. If P is in one of the subsets and Q in the other, then $\overline{PQ}$ intersects ℓ [as pictured in (b) in the following figure].

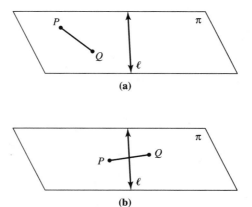

Each of the subsets above is called a half plane. If P is a point of π in one of the subsets, then we refer to the half plane containing P as "the P side of ℓ" and denote it as $\pi_{\text{half}}(P, \ell)$, as shown in the following figure .

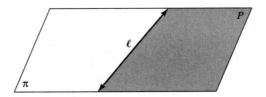

- The two rays in part (a) below are called parallel rays. They lie in the same half plane.

The two rays in part (*b*) below are called antiparallel rays. They lie in different half planes.

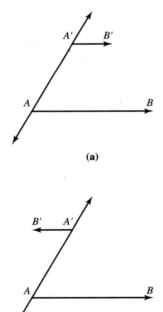

(a)

(b)

If $\overrightarrow{AB} \parallel \overrightarrow{A'B'}$ and $\overrightarrow{AC} \parallel \overrightarrow{A'C'}$, then $\angle CAB = \angle C'A'B'$, where these angles are measured in $\pi(A, B, C)$ and $\pi(A', B', C')$ respectively, as shown in the following figure .

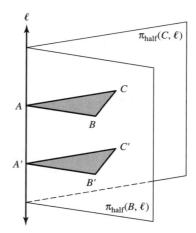

- A dihedral angle is a figure in space formed by two half planes with a common boundary line ℓ, called the edge. Each of the half planes is referred to as a face of the dihedral angle. $\angle BAC$ is the plane angle of the dihedral angle pictured above. All such plane angles are congruent as well. Also, vertical dihedral angles formed by intersecting plane angles are congruent.

- If two planes intersect and any one of the dihedral angles formed measures 90°, then each of the four dihedral angles formed is also 90°.

- If π_1 and π_2 are planes and ℓ is a line contained in π_2 with $\ell \perp \pi_1$, then $\pi_1 \perp \pi_2$.

- If π is a plane and P is a point not on π, then there is a unique line through P perpendicular to π. Let F be the point where the perpendicular intersects π, and we write $F = \mathrm{proj}_\pi(P)$, meaning F is the projection of P on π. If P is already on π, then $\mathrm{proj}_\pi(P) = P$. If $\ell \perp \pi$, then $\mathrm{proj}_\pi(\ell)$ is the single point at which ℓ intersects π.

  If π is a plane and ℓ a line not perpendicular to π, then $\mathrm{proj}_\pi(\ell)$ is a line contained in π, as shown in the next figure.

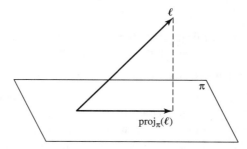

- If π is a plane and ℓ a line not perpendicular to π, then there is a unique plane π' containing ℓ such that $\pi' \perp \pi$, as shown below.

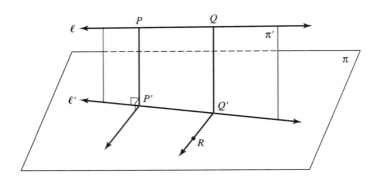

- Let ℓ be a line which intersects the plane π at A with ℓ not perpendicular to π. Then the acute angle between ℓ and $\text{proj}_\pi(\ell)$ is less than the angle between ℓ and any other line of π passing through A. (This is known as the angle between a line and a plane.)

- If ℓ and ℓ' are skew lines, then there are points P on ℓ and P' on ℓ' which have the following three properties:

  1. $\overleftrightarrow{PP'} \perp \ell$ and $\overleftrightarrow{PP'} \perp \ell'$.
  2. PP' is the unique line satisfying 1 above.
  3. $\overline{PP'}$ is the shortest segment from a point on ℓ to a point on ℓ'.

- The trihedral angle pictured below in part (a), denoted $\angle AB_1B_2B_3$, is formed when the edges of three half planes meet in a common point, the vertex, A (also denoted first before the edge points in the notation). The edges are $\overrightarrow{AB_1}$, $\overrightarrow{AB_2}$ and $\overrightarrow{AB_3}$. The faces are composed of points of the planes $\pi(A, B_1, B_2)$, $\pi(A, B_2, B_3)$ and $\pi(A, B_3, B_1)$ bounded by the edges. The three angles $\angle B_1AB_2$, $\angle B_2AB_3$, and $\angle B_3AB_1$ are the face angles. The trihedral angle shown is composed of portions of three dihedral angles: the dihedral angle between $\pi_{\text{half}}(\overleftrightarrow{AB_1}, B_2)$ and $\pi_{\text{half}}(\overleftrightarrow{AB_1}, B_3)$, between $\pi_{\text{half}}(\overleftrightarrow{AB_2}, B_1)$ and $\pi_{\text{half}}(\overleftrightarrow{AB_2}, B_3)$, and between $\pi_{\text{half}}(\overleftrightarrow{AB_3}, B_2)$ and $\pi_{\text{half}}(\overleftrightarrow{AB_3}, B_1)$.

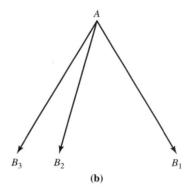

- Given triangles $\triangle ABC$ and $\triangle A'B'C'$, if $\overline{AB} \cong \overline{A'B'}$ and $\overline{AC} \cong \overline{A'C'}$, then $\angle A > \angle A'$ if and only if $BC > B'C'$.

- The sum of any two face angles of a trihedral angle is greater than the third face angle.

- The sum of the face angles of a trihedral angle is less than $360°$.

- Two trihedral angles are congruent or symmetric if the following six conditions hold:

   1. $\angle B_1 A B_2 \cong \angle B_1' A' B_2'$

   2. $\angle B_2 A B_3 \cong \angle B_2' A' B_3'$

   3. $\angle B_3 A B_1 \cong \angle B_3' A' B_1'$

   4. the dihedral angle with edge $\overrightarrow{AB_1} \cong$ the dihedral angle with edge $\overrightarrow{A'B_1'}$

   5. the dihedral angle with edge $\overrightarrow{AB_2} \cong$ the dihedral angle with edge $\overrightarrow{A'B_2'}$

6. the dihedral angle with edge $\overrightarrow{AB_3} \cong$ the dihedral angle with edge $\overrightarrow{AB'_3}$

Trihedral angles are congruent if face angles and dihedral angles of one are equal to corresponding face angles and dihedral angles of the other when arranged in the same order. If arranged in the opposite order, then the two trihedral angles are mirror images of each other, and we say they are symmetric. (See part (b) of picture on previous figure.)

- If the three face angles of one trihedral angle are congruent respectively to the three face angles of another trihedral angle, then the two trihedral angles are either congruent or symmetric.

CHAPTER 13

Combinatorial Theorems in Geometry

13.1 THE TRIANGULATION LEMMA

Combinatorics is often described as the branch of mathematics that deals with problems of counting, although the definition is probably not too helpful if you are not experienced with the subject. The main results of this chapter compare the numbers of edges, sides, and faces in various geometric configurations. We are including this chapter here because of Euler's theorem on polyhedra. Our treatment is mainly based on the article "The equivalence of Euler's and Pick's theorem," by D. DeTemple and J. M. Robertson, in *Mathematics Teacher*, 67(March 1974), pp. 222–226.

At any rate, this section is devoted to triangulations of polygons. A *triangulation* of a polygon is a way of cutting the polygon into nonoverlapping triangles such that if two of the triangles touch, the intersection is either a vertex of each or an edge of each (Fig. 13.1). Given such a triangulation, let I be the number of interior vertices, B be the number of vertices of the polygon (boundary vertices), and F be the number of triangles (faces). So, in the figure, $I =?$, $B =?$, and $F =?$.

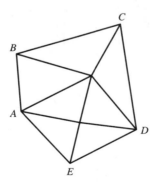

Figure 13.1: Triangulation of a polygon

THE TRIANGULATION LEMMA. $F = 2I + B - 2$.

Proof. Let S be the sum of all the angles of the triangles. We will compute S two different ways. On the one hand, since there are F triangles and each triangle's angles add to 180°,

$$S = F \cdot 180°.$$

On the other hand, instead of organizing the angles according to which triangle they belong to we can organize them by vertex. The sum of the angles at each interior vertex is 360°. At each boundary vertex the sum of the angles is equal to one of the angles of the polygon, so their sum is equal to the sum of the angles of the polygon. Since the number of sides of the polygon equals the number of vertices, which is B, this sum is $(B - 2)180°$. Adding gives

$$S = 360° \cdot I + 180°(B - 2).$$

Setting these two expressions for S equal and dividing by 180° gives the desired formula. ☐

13.2 EULER'S THEOREM

Which combinations of plane polygons can be put together to make a polyhedron in space and how many different ways are there to do this? This is a difficult question and part of the answer is given by Euler's theorem. Euler's theorem relates the number of faces, edges and vertices in a polyhedron. Before stating the theorem, let's work out some examples.

One type of polyhedron we mentioned in Chapter 12 was the tetrahedron. Each tetrahedron has four vertices, six edges and four faces. A generalization of the tetrahedron is the pyramid (Fig. 13.2). A pyramid has a base consisting of a planar polygon and a vertex not in the plane with edges connecting it to each vertex of the base. If the polygon has n sides, then the pyramid has $n + 1$ vertices, $2n$ edges, and $n + 1$ faces. Another generalization is the double pyramid. It is constructed from a planar polygon by adding two vertices (separated by the plane) with each vertex of the polygon joined to both of these vertices. If the polygon has n sides, then this double pyramid will have $n + 2$ vertices, $3n$ edges, and $2n$ faces.

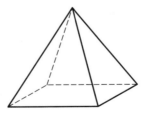

Figure 13.2: Pyramid

Perhaps the most familiar polyhedron is the cube. A cube has 8 vertices, 12 edges, and 6 faces. One generalization of the cube is called a cylindrical solid. It is formed by two congruent polygons in parallel planes, with corresponding sides parallel and corresponding vertices joined (Fig. 13.3). If the polygon has n sides, then the cylindrical solid has $2n$ vertices, $3n$ edges, and $n + 2$ faces.

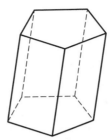

Figure 13.3: Cylindrical solid

What is the pattern? In every case we considered the number of vertices and faces together was 2 more than the number of edges. You are encouraged to check this and to try additional polyhedra. At any rate, it does always work, and this is Euler's theorem.

EULER'S THEOREM. Given any polyhedron, let V be the number of vertices, E the number of edges, and F the number of faces. Then $V - E + F = 2$.

We will prove Euler's theorem using the triangulation lemma. This will involve two reductions: One to plane figures and one to triangulations.

Reduction to two dimensions. Given a polyhedron, if we remove one of the faces, then, at the cost of some stretching, we can flatten the rest into a plane figure. In Fig. 13.4, we show how this construction works for a square pyramid, removing the square base or removing one of the triangular sides.

The resulting figure is a polygon broken into a collection of nonoverlapping polygons such that, if any two of them touch, then the intersection is either a vertex of each or an edge of each. We call such a collection of polygons a *polygonal tiling* of the big polygon. This generalizes the concept of triangularization. □

Given our reduction, we now need to prove the following version of Euler's theorem.

THEOREM. Given any polygonal tiling of a polygon, let V be the number of vertices, E the number of edges, and F the number of faces. Then

$$V - E + F = 1.$$

Reduction to triangular faces. Given any polygonal tiling of a polygon T define $\chi(T) = V - E + F$, where V is the number of vertices, E is the number of edges, and F is the number of faces. We want to prove that $\chi(T)$ is always 1, and we want to prove it using the triangulation lemma. In this step, we show that if T is any polygonal tiling, then we can find a polygonal tiling S in which every face is a triangle such that $\chi(T) = \chi(S)$.

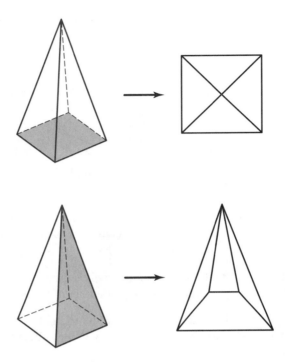

Figure 13.4: Flattening a square pyramid

If some face of T is not a triangle, we may construct a new tiling T' by adding in one diagonal. The effect is to keep the number of vertices the same, increase the number of edges by 1, and increase the number of faces by 1. Hence, $\chi(G) = \chi(T')$. If we repeat this operation enough times, we obtain a tiling S in which every face is a triangle and in which $\chi(T) = \chi(S)$, as claimed. $\square$

Proof of Euler's theorem. Given any tiling T, we construct S as before. We now use the triangulation lemma to prove that $\chi(S) = 1$. By the definition of S this will imply that $\chi(T) = 1$, and by the first reduction this implies Euler's theorem.

In order to apply the triangulation lemma we need to relate the numbers I and B to the numbers V, E, and F. One obvious relation is

$$I + B = V.$$

To get the other relation we need we consider the set of all pairs (e, f) where e is an edge, f is a face, and e is on f. How many such pairs are there? On the one hand, since every face is a triangle, every face has three edges and so there are $3F$ such pairs. On the other hand, internal edges are on two faces and edges on the boundary are on one edge. It follows that the number of such pairs is $2(E - B) + B = 2E - B$ and so

$$3F = 2E - B.$$

Hence, $I = V - B = V - (2E - 3F)$. Finally, we can substitute this information into the triangulation lemma to get

$$F = 2I + B - 2 = V + V - 2E + 3F - 2.$$

The rest of the proof follows from an easy algebraic computation. $\qquad\square$

BIG ASIDE. In two dimensions, a polygon has vertices and edges and their number is related by $V = E$, or $V - E = 0$. In three dimensions a polyhedron has vertices, edges, and faces, and Euler's theorem says that $V - E + F = 2$. If we go to four-dimensions (whatever that is), a four dimensional "hyperpolyhedron" has vertices (points), edges (line segments), faces (polygons), and hyperfaces (polyhedra). If the number of each is V, E, F, and H it is reasonable to expect that $V - E + F - H$ will be a constant. It is, and the constant is zero, $V - E + F - H = 0$. More generally, in n dimensions the "Euler-Poincaré characteristic" will be either 0 or 2, depending on whether the dimension n is even or odd.

13.3 PLATONIC SOLIDS

DEFINITION. A polyhedron is a Platonic solid if all the faces are congruent regular polygons and every vertex is on an equal number of sides.

The most familiar example of a Platonic solid is a cube. See Exercises 1 and 2 for Platonic solids that are triangular pyramids and square double pyramids. The surprising fact is that there are only five types of Platonic solids. This is in stark contrast to plane geometry, where, if we consider the two-dimensional analogue of Platonic solids to be regular polygons, there are regular n-gons for every $n \geq 3$.

LEMMA. If a Platonic solid has n edges in every face and d edges at every vertex, then $2E = nF = dV$.

Proof. We first prove that $2E = nF$. The basic idea is that every edge is on two faces and every face has n edges. More formally, consider the set of pairs (e, f) in which e is an edge, f is a face, and e is on f. We count the number of such pairs in two different ways: On the one hand, there are two faces containing each edge e, so the number of pairs is $2E$. On the other hand, there are n edges in each face, so the number is also nF. So, $2E = nF$.

The proof of $2E = dV$ is similar, using the pairs (v, e) in which v is a vertex, e is an edge and v is an endpoint of e. $\qquad\square$

LEMMA. For n and d as before, $2n + 2d > nd$.

Proof. In Euler's theorem, $V - E + F = 2$, we set $V = 2E/d$ and $F = 2E/n$. This gives

$$(\frac{2}{d} - 1 + \frac{2}{n})E = 2.$$

It follows that

$$\frac{2}{d} - 1 + \frac{2}{n} > 0.$$

To complete the proof, just multiply both sides by nd. $\square$

THEOREM. In any Platonic solid, the faces are either equilateral triangles, squares or regular pentagons. If the faces are triangles, then each vertex has either three, four, or five edges. If the faces are squares or pentagons, then each vertex has three edges.

Proof. In the previous lemma, set $n = 3$ to get

$$6 + 2d > 3d.$$

This implies that $d < 6$, so d must be 3, or 4 or 5. Next, if $n = 4$, then

$$8 + 2d > 4d.$$

Hence, $d < 4$, and so $d = 3$. Likewise, if $n = 5$, then

$$10 + 2d > 5d.$$

This implies that $10 > 3d$ and, since d is an integer at least 3, we must have $d = 3$. Finally,

$$2n + 2d - nd = 2n - (n - 2)d.$$

Since $d \geq 3$ this expression is at most $2n - (n-2)3$, because increasing d would cause a larger number to be subtracted. But

$$2n - (n - 2)3 = 6 - n > 0,$$

and so $n < 6$. This completes the proof. $\square$

Technically, this theorem only proves that there are at most five types of Platonic solids, because it only shows that there are five sets of numbers that are consistent with Euler's theorem. This is only algebra. To show that all five actually exist we would need to do geometry. At any rate, all five do exist. This would be a good time for you to cut out a number of congruent equilateral triangles and try to tape them together to make polyhedra with three, four, or five triangles at each vertex.

Using the equations $2E = nF$, $2E = dV$, and $E = 1/(\frac{2}{n} - 1 + \frac{2}{d})$, we can solve for V, E, and F in each case. If $n = d = 3$, then $E = 6$ and $V = F = 4$. The polyhedron would be a triangular pyramid and it is called a regular tetrahedron. If $n = 3$ and $d = 4$, then $E = 12$, $V = 6$, and $F = 8$. The polyhedron is a square double pyramid and it is called a regular octahedron. If $n = 3$ and $d = 5$ then $E = 15$, $V = 6$, and $F = 10$. This polyhedron is called an isocohedron. We do not describe how it is constructed. If $n = 4$ the faces are squares and, of course, the polyhedron is a cube. Finally, if $n = 5$ and $d = 3$, then $E = 15$, $V = 10$, and $F = 6$. The polyhedron is called a dodecahedron and if you take a careful look at a soccer ball you may see how it is constructed.

13.4 PICK'S THEOREM

Pick's theorem is a theorem in plane geometry and it calculates the areas of polygons whose vertices are lattice points. In a coordinate system lattice points are points whose coordinates (x, y) are both integers. (Fig. 13.5). It is reasonable to expect that if a polygon has a large area, then it would contain a large number of lattice points. Given a polygon with lattice points for vertices, let B be the number of lattice points on the boundary and let I be the number of lattice points in the interior.

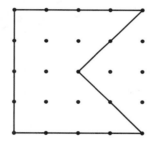

Figure 13.5: Lattice polygon

PICK'S THEOREM. For any polygon whose vertices are lattice points, the area is $I + \frac{1}{2}B - 1$.

So, for example, in the 1×1 square, $I = 0$, $B = 4$, and, of course the area is 1, which equals $0 + 4/2 - 1$. And in the 2×2 square, $I = 1$, $B = 8$, and the area is 4, which equals $1 + 8/2 - 1$. Of course, part of the power of Pick's theorem and one reason it is not trivial to prove is that it is true for any polygon, not just for ones with horizontal and vertical sides.

Pick's theorem suggests that the smallest polygon with lattice points for vertices would be a triangle with $B = 3$ and $I = 0$ and that the area of such a figure would be $\frac{1}{2}$. We call a triangle in which $B = 3$ and $I = 0$ a *fundamental triangle*, (Fig. 13.6). Given any polygon we may triangulate it (perhaps in more than one way) using only fundamental triangles. See Fig. 13.7. The triangulation lemma now says that

$$F = 2I + B - 2.$$

To complete the proof, we need to prove that every fundamental triangle has area $\frac{1}{2}$ because that will imply that F is one-half the area of the polygon. So, dividing the equation by 2 gives that area is $I + \frac{1}{2}B - 1$, as claimed.

Figure 13.6: Fundamental triangle

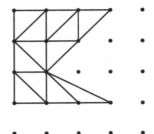

Figure 13.7: Lattice polygon triangulated by fundamental triangles

LEMMA. Every fundamental triangle has area $\frac{1}{2}$.

In order to prove this lemma we first need to prove a technical result.

SUBLEMMA. Let $\triangle ABC$ be a fundamental angle with largest angle at B. Then $\angle B \geq 90°$. Moreover, if $\triangle ABC$ is a right triangle, then the legs have length one and are parallel to the X- and Y-axes.

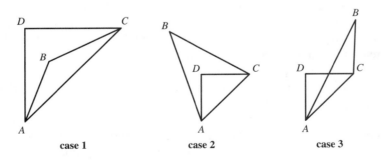

Figure 13.8: Proof of sublemma

Proof of Sublemma. Let $\triangle ABC$ be a fundamental triangle with largest angle $\angle B$. Then the largest side will be $\overline{AC}$. For convienience, we will assume that

$\overline{AC}$ has positive slope, A is below C, and B is above $\overline{AC}$. After reading the proof you may wish to supply proofs in the other possible cases. They are all very similar.

Let us use a coordinate system with origin at $A = (0, 0)$ and let $C = (c_1, c_2)$ (the coordinates will be positive) and $B = (b_1, b_2)$. Let D be the point $(0, c_2)$ so that $\triangle ADC$ will be a right triangle. We now compare the angles of $\triangle ABC$ with those of $\triangle ADC$ at the vertices A and C. There are three possibilities. First, if $\angle BAC \leq \angle DAC$ and $\angle BCA \leq \angle DCA$, then B is inside of $\triangle ADC$. In this case,

$$\angle B = 180° - (\angle BAC + \angle BCA) \geq 180° - (\angle DAC + \angle DCA) = \angle D.$$

So $\angle B \geq 90°$ and we are done in this case. Also note that the only way that $\angle B$ could be a right angle would be if $B = D$. The rest of the proof will consist of showing that the remaining cases are impossible.

If $\angle BAC \geq \angle DAC$ and $\angle BCA \geq \angle DCA$ with at least one of the inequalities being strict, then the point D would be on a side or in the interior of $\triangle ABC$ and this would contradict the assumption that $\triangle ABC$ is fundamental. This leaves only the possibility that one of the angles from $\triangle ABC$ is strictly less than the angle coming from $\triangle ADC$ and the other is greater than or equal. We will do the case of $\angle BAC < \angle DAC$ and $\angle BCA \geq \angle DCA$. In this case, the line segment $\overline{AB}$ crosses the line segment $\overline{CD}$. Consider the point $E = (c_1 - 1, c_2)$. Since $\triangle ABC$ is fundamental, E cannot be an interior point and so $\overline{AB}$ must pass between E and C. This implies that $b_1 > c_1$ and $b_2 \geq c_2$. But it now follows easily from the distance formula from analytic geometry (= the Pythagorean theorem) that $\overline{AB}$ is longer than $\overline{AC}$. This contradicts our assumpti   □

on that $\overline{AC}$ is the longest side and so completes the proof.

Proof of lemma. Given a fundamental triangle $\triangle ABC$ with largest angle $\angle B$. We will show that if $\angle B$ is not a right angle we can construct a fundamental triangle DBC with the same area as $\triangle ABC$ and with $DC = AB$ and $DB < AC$. To do this, choose D so that $ABCD$ is a parallelogram. It is not hard to see that $\triangle ADC$ is also a fundamental triangle and so $ABCD$ has four lattice points on the boundary and none in the interior. The diagonal of a parallelogram divides it into two congruent triangles. In particular, each of the triangles has area equal to half of the area of the parallelogram. So $\triangle ABD$ and $\triangle DBC$ are fundamental triangles with the same area. Also, $\overline{DC} \cong \overline{AB}$, since they are opposite sides of a parallelogram. To see that $DB < AC$, use the scissors lemma on the triangles $\triangle BCD$ and $\triangle ABC$, using the fact that $\angle B$ is obtuse and so is greater than $\angle C$, which is acute. This completes the proof of the construction.

So, given any fundamental triangle T we may do this construction repeatedly to get a sequence of triangles $T_1, T_2, \ldots$. In each step the length of the longest side decreases. But, the distance between any two lattice points is the square root

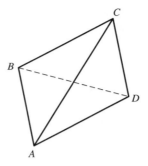

Figure 13.9: Proof that fundamental triangle has area $\frac{1}{2}$

of an integer. So after some finite number of steps the process must terminate with a right triangle whose area is $\frac{1}{2}$.   □

PROBLEMS

13.1. Let $\triangle ABC$ be an equilateral triangle with center O. Let ℓ be the line through ℓ perpendicular to the plane of $\triangle ABC$. Prove that there is a point D on ℓ such that the triangles $\triangle DAB$, $\triangle DBC$, $\triangle DCB$, and $\triangle ABC$ are all congruent. Calculate the length OD in terms of AB.

13.2. Let $ABCD$ be a square with center O. Let ℓ be the line through ℓ perpendicular to the plane of $ABCD$. Prove that there is a point E on ℓ such that the four triangles $\triangle EAB$, $\triangle EBC$, $\triangle ECD$, and $\triangle EDA$ are all equilateral. Calculate the length OD in terms of AB. How can you use this construction to construct a regular octahedron?

13.3. If you imitate the construction of the previous two exercises using a regular pentagon as a base, why can't you get a construction of a Platonic solid?

13.4. Given a Platonic solid P you can construct a new polyhedron whose vertices are the centers of the faces of P. This new polyhedron is called the dual of P and it turns out that it is also a Platonic solid. For each of the five types of Platonic solids, identify the dual.

13.5. Our proof of Euler's theorem involved "flattening" a polyhedron. Draw the flattened form of a regular tetrahedron, a cube, and a regular octahedron.

13.6. Use Pick's theorem to calculate the area of each of these figures.

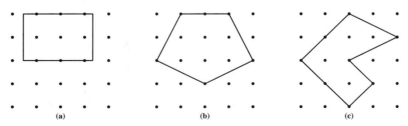

13.7. Calculate I, B, and area for Fig. 13.10. Why does this not contradict Pick's theorem?

13.8. Prove that if A and B are lattice points such that there are no lattice points on $\overline{AB}$ and C is a lattice point that minimizes the distance to $\overleftrightarrow{AB}$, then $\triangle ABC$ is fundamental.

13.9. Technically, Euler's theorem only holds for polyhedra with no "holes" in them (genus 0). Verify that in the following polyhedron $V = 16$, $E = 32$ and $F = 16$ so $V - E + F = 0$. Can you see where the proof of Euler's theorem uses this assumption?

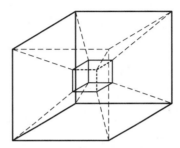

13.10. Given $a > c > 0$ and $b > d > 0$ define $\triangle ABC$ to be the triangle with vertices $A = (0,0)$, $B = (a, b)$ and $C = (c, d)$. Prove that $\triangle ABC$ has area $\frac{1}{2}(bc - ad)$. Use this to find a point C such that $\triangle ABC$ will be fundamental, for $B = (5, 3)$.

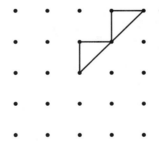

Figure 13.10: Exercise 13.7

CHAPTER SUMMARY

- Combinatorics is a branch of mathematics which deals with counting problems. Relations between counts of edges, sides and faces in various geometric configurations are studied in this chapter.

- A triangulation of a polygon in the plane is a way of sectioning it into non-overlapping triangles whose only intersections are either vertices or edges of each. Given a triangulation, let F be the number of faces (triangles), I be the number of interior vertices, and let B be the number of vertices on the boundary of the polygon.

  Triangulation Lemma. $F = 2I + B - 2$.

- A solid in space bounded by planes is a polyhedron each planar face of which is a plane polygon called a face of the polyhedron. Let the number of boundary vertices of the polyhedron be V and the number of boundary edges be E.

  Euler's Theorem. $V - E + F = 2$.

- A polyhedron is a Platonic solid if all its faces are congruent regular polygons and every vertex is on an equal number of sides.

  In a Platonic solid, the faces are either equilateral triangles, squares or regular pentagons. If the faces are triangles, then each vertex has either 3, 4 or 5 edges. If the faces are squares or pentagons, then each vertex has 3 edges.

- There are five Platonic solids: the regular tetrahedron, the regular octahedron, the cube, the icosohedron, and the dodecahedron.

- Lattice points are points whose coordinates with respect to a set of coordinate axes are integer pairs.

- Pick's Theorem. A polygon in the plane whose vertices are lattice points has area = I + (1/2)B - 1 , where B is the number of lattice points on the boundary and I is the number of lattice points in the interior of the polygon.

CHAPTER 14

Spherical Geometry

14.1 SPHERES AND GREAT CIRCLES

The surface of a sphere is just as handy a model for practical problems of a global nature as the plane is for local problems. However we do not study the sphere (meaning its surface) only because it is the natural idealized model of earth. We study it also because it will provide a basis for us to model a geometry which spectacularly disobeys one of Euclid's axioms, the fifth, or parallel postulate. We have already considered the Playfair formulation of Euclid's (P5), namely, there is one and only one parallel to a given line through a point not on the line. As we shall see in the last section of this chapter, with the proper interpretation of the words "points"and "lines,"we can provide a model for a geometry that is non-Euclidean in the sense that it will obey most of Euclid's axioms but will have no parallels.

We take a sphere to be the set of points in space at a given distance from a fixed point. We call the fixed point the center of the sphere. Define a radius of the sphere as any segment from the center to a point on the sphere so that the length of any radius is the given distance. A diameter is a segment through the center whose end points are on the sphere and whose length is therefore twice that of a radius. A sphere is taken to be uniquely specified once its center and the length of a radius are given.

The first result we study is concerned with a natural way to produce circles on spheres. Recall that the notation $dist(P, \pi)$ indicates the distance from a point P to a plane π and is discussed in Chapter 12. Also, we adopt the convention of referring to "a sphere of radius r"rather than "a sphere whose radii all have length r."

> **THEOREM.** If a plane π intersects a sphere with center O and radius r a distance $x = dist(O, \pi)$ from the center, where $0 \leq x < r$, then the intersection is a circle with radius $\sqrt{r^2 - x^2}$.

> ***Proof.*** Let $O' = \text{proj}_\pi(O)$, the foot of the perpendicular from O to π. If P and Q are any points in the intersection, then $OP = OQ$ since P and Q are on the sphere (see Fig. 14.1) and $\overline{OO'}$ is perpendicular to each of $\overline{O'P}$ and $\overline{O'Q}$ since P and Q are on π. By the hypotenuse-leg congruence theorem, $\triangle OO'P \cong \triangle OO'Q$, and we have $O'P = O'Q$. Thus, any point of the intersection is the same fixed distance from O' and therefore the intersection is a circle. From the theorem of Pythagoras, $O'Q$ is the length of any radius and equals $\sqrt{r^2 - x^2}$. $\square$

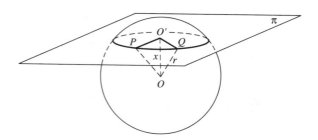

Figure 14.1: Intersection of sphere and plane

When $x = 0$ in the preceding theorem, the plane passes through the center of the sphere and the ensuing circle is called a *great circle;* the radius of a great circle clearly equals the radius of the sphere. We refer to the end points of a diameter of a great circle (which is also a diameter of the sphere) as *antipodes.* Any other circle formed by intersecting a plane with a sphere is called a *small circle* of the sphere. Given any circle of a sphere, the spherical diameter perpendicular to the plane of this circle passes through its center (by the proof of the preceding theorem) and is called the *axis* of the circle. Finally, the end points of the axis of any circle are called its *poles.*

COROLLARY. A unique great circle is determined by any two points of a sphere that are not antipodes.

Proof. Let O be the center of the sphere and P and Q be the two points. If O, P, and Q were collinear, then P and Q would be antipodes, which is not the case. So the three points determine a unique plane $\pi(O, P, Q)$ which intersects the sphere in a unique great circle.   □

COROLLARY. Three points on a sphere determine a unique circle of the sphere.

Proof. The three points determine a unique plane, which in turn intersects in the circle.   □

14.2 SPHERICAL TRIANGLES

We will become used to thinking of great circles on a sphere as analogous to lines in the plane and to arcs of great circles as analogous to segments. In this spirit, we define a *spherical angle* as an angle on the sphere formed by two intersecting arcs of great circles. The planes of the two great circles containing these arcs each pass through the center of the sphere and thus intersect in a line containing the common diameter of the circles. Thus, in Fig. 14.2, the arcs $\overset{\frown}{PA}$ and $\overset{\frown}{PB}$ (P, A and P, B are assumed to not be pairs of antipodes) intersect at P and determine great circles and unique planes whose intersection contains the diameter $\overline{PQ}$. The two half planes $\pi_{\text{half}}(A, \overleftrightarrow{PQ})$ and $\pi_{\text{half}}(B, \overleftrightarrow{PQ})$ determine a dihedral angle (see Chapter 12) with edge $\overleftrightarrow{PQ}$. In general, the arcs of any spherical angle thus determine unique planes that intersect in a unique

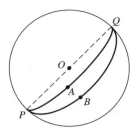

Figure 14.2: Spherical angle $\angle APB$

dihedral angle containing these arcs, and we define the degree measure (or size) of the spherical angle to be the degree measure of the corresponding dihedral angle. The next theorem shows that there is a way to measure the size of a spherical angle along an arc. As usual, we use the same notation for both a spherical angle and its measure.

THEOREM. Let $\angle APB$ be a spherical angle, and suppose C is the unique great circle that has P as one of its poles. Extend the arcs $\overarc{PA}$ and $\overarc{PB}$ so as to intersect C at A' and B', respectively. Then $\angle APB = \overarc{A'B'}$, where $\overarc{A'B'}$ is measured along C.

Proof. If Q is the antipode of P, then we may easily use two radii of the sphere that are perpendicular to $\overline{PQ}$ at the center to determine the unique plane perpendicular to $\overline{PQ}$. The intersection of this plane with the sphere is C, whose poles are P and Q by construction. Now suppose the great circle determined by P and A intersects C at A' and that the great circle determined by P and B intersects C at B' (see Fig. 14.3). Since $\overline{OP}$ is perpendicular to the plane of C

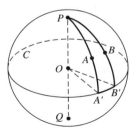

Figure 14.3: Spherical $\angle APB = \overarc{A'B'}$

at O, $\overline{OP} \perp \overline{OA'}$ and $\overline{OP} \perp \overline{OB'}$. Consequently, $\angle A'OB'$ is a plane angle of the dihedral angle determined by the spherical angle $\angle APB$. Because a dihedral angle is measured by any of its plane angles and $\angle A'OB'$ is the central angle of $\overarc{A'B'}$, it follows that $\angle APB = \overarc{A'B'}$.   □

We now define a *spherical triangle* to be the closed figure formed on a sphere by three arcs of great circles which intersect at their end points. The arcs of the circles are sides of the spherical triangle and the three intersection points of the sides are called vertices. The angles of a spherical triangle are the spherical angles formed by the arcs. Thus, in Fig. 14.4, $\triangle ABC$ is a spherical triangle; its sides are the arcs $\overparen{AB}$, $\overparen{AC}$; and $\overparen{BC}$, and its angles are the spherical angles $\angle ABC$, $\angle BCA$, and $\angle CAB$. Naturally, points A, B, and C of the sphere are the vertices of $\triangle ABC$.

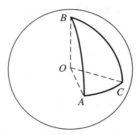

Figure 14.4: A spherical triangle and its associated trihedral angle

Given any spherical $\triangle ABC$, we can construct an associated trihedral angle (Chapter 12). Specifically, we connect the center of the sphere to each of the three vertices by forming the radii $\overline{AB}$, $\overline{OB}$, and $\overline{OC}$. The three planes $\pi(O, A, B)$, $\pi(O, A, C)$, and $\pi(O, C, B)$ all intersect at point O to form the trihedral angle $\angle OBAC$. The spherical $\triangle ABC$ is the intersection of $\angle OBAC$ with the sphere. The sides of the spherical triangle are measured by the face angles of the trihedral angle; namely, $\angle BOA = \overparen{BA}$, $\angle BOC = \overparen{BC}$, and $\angle AOC = \overparen{AC}$. Moreover, the spherical angles of $\triangle ABC$ are measured by the corresponding dihedral angles of the trihedral angle. We can immediately make use of the fact that a trihedral angle is associated to a given spherical triangle to get quick proofs of the next two theorems.

THEOREM. The sum of any two sides of a spherical triangle is greater than the third. This is the triangle inequality for spherical triangles.

Proof. This result is equivalent to the trihedral result of Chapter 12 that the sum of any two face angles of a trihedral angle exceeds the third.   □

THEOREM. The sum of the sides of a spherical triangle is less than $360°$.

Proof. Here, the corresponding trihedral result is that the sum of the face angles of a trihedral angle is less than $360°$.

□

Two points on a sphere that are not antipodal divide the great circle they determine into the union of two arcs. We call the shorter of these two arcs (why must one be shorter?) the *minor arc* of the great circle joining these points. The distance between two such points is taken to be the length (spherical) of the minor arc joining them. We can show that this minor arc is the shortest polygonal path (finite sequence of great circle arcs connected end to end) connecting two points.

COROLLARY. The distance between two non-antipodal points on a sphere is shorter than the length of any polygonal path between these points.

Proof. Let the points be A and B with $\overset{\frown}{AB}$ the minor arc between them. Suppose they are joined by the path $\overset{\frown}{AC}$ and $\overset{\frown}{CB}$, where C is not on $\overset{\frown}{AB}$. We have $\overset{\frown}{AB} < \overset{\frown}{AC} + \overset{\frown}{CB}$ by the triangle inequality applied to the spherical $\triangle ABC$. Next, suppose A and B are joined by the path $ACDB$ for C and D not on $\overset{\frown}{AB}$ (Fig. 14.5). In spherical $\triangle ABC$, $\overset{\frown}{AB} < \overset{\frown}{AC} + \overset{\frown}{CB}$. Also, in the spherical

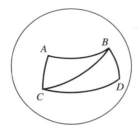

Figure 14.5: Polygonal path

$\triangle BCD$,

$$\overset{\frown}{CB} < \overset{\frown}{CD} + \overset{\frown}{DB}\ .$$

By substitution,

$$\overset{\frown}{AB} < \overset{\frown}{AC} + \overset{\frown}{CB} < \overset{\frown}{AC} + \overset{\frown}{CD} + \overset{\frown}{DB},$$

as desired. A similar argument will work for any path with any finite number of arcs. If you are familiar with mathematical induction, you may enjoy writing out a more formal induction proof. □

We remark that the minor arc joining two points is shorter than any other continuous curve on the sphere between these two points. The proof of this requires calculus-based methods.

14.3 POLAR TRIANGLES

We start with a useful lemma concerning the poles of a great circle.

LEMMA. The spherical distance between either pole of a great circle and any point of the great circle is 90°.

Proof. Denote the center of the sphere by O, two points on the great circle by A and B, and the poles of the great circle by P and Q, as in Fig. 14.6. By definition

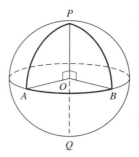

Figure 14.6: Polar distance $= 90°$

of the term "pole," $\overleftrightarrow{PQ}$ is perpendicular to the plane of the great circle at O. The radii $\overline{OA}$ and $\overline{OB}$ are in this plane and therefore $\angle POA = \angle POB = 90°$, which in turn implies that $\overparen{PA} = \overparen{PB} = 90°$, because these arcs are in the planes $\pi(P, O, A)$ and $\pi(P, O, B)$, respectively. However, $\overparen{PA}$ and $\overparen{PB}$ are the minor arcs joining P to the arbitrary points A and B of the great circle, so the distance of P to any point of the great circle is $90°$. Since the same reasoning holds for Q, the lemma is proved. $\square$

Spherical $\triangle A'B'C'$ is called the *polar triangle* of spherical $\triangle ABC$ if A is the pole on the A' side of the great circle containing $\overparen{B'C'}$, B is the pole on the B' side of the great circle of $\overparen{A'C'}$, and C is the pole on the C' side of the great circle of $\overparen{A'B'}$ (Fig. 14.7). Our first result on polar triangles shows that this definition is symmetric.

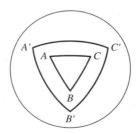

Figure 14.7: Polar triangles

THEOREM. If spherical $\triangle A'B'C'$ is the polar triangle of spherical $\triangle ABC$, then $\triangle ABC$ is the polar triangle of $\triangle A'B'C'$.

Proof. We must show that the vertices of $\triangle A'B'C'$ are poles of the sides of $\triangle ABC$. We will just show that A' is a pole of $\overset{\frown}{BC}$ since the details showing that B' and C' are the poles of $\overset{\frown}{AC}$ and $\overset{\frown}{AB}$ proceed in a like manner.

From the definition of polar triangle, B is a pole of the great circle of $\overset{\frown}{A'C'}$. Applying our introductory lemma, we see that the distance of A' from B is $90°$. Similarly, C is a pole of $\overset{\frown}{A'B'}$, so the distance of A' from C is $90°$. We now claim that A' must be a pole of the great circle containing $\overset{\frown}{BC}$ (Fig. 14.8).

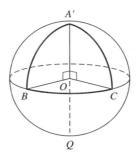

Figure 14.8: Two quadrants and a pole

To see this, construct $\overline{A'O}$, $\overline{OB}$, and $\overline{OC}$ and note that what we have said about distances means $\angle A'OB = \angle A'OC = 90°$. Thus, $\overleftrightarrow{A'O}$ is perpendicular to the plane of the great circle of $\overset{\frown}{BC}$ at O. Since A' is on the sphere it must be the end point of an axis and is therefore a pole of $\overset{\frown}{BC}$. Moreover, if A' were not on the A side of the great circle of $\overset{\frown}{BC}$, then A would not be on the A' side of the great circle of $\overset{\frown}{B'C'}$. This establishes the claim and the theorem. $\square$

We can now play off the symmetry of the polar triangles to obtain detailed angle relationships. In particular, we show that each angle of one triangle of a polar triangle pair is the supplement of the opposite side of the other triangle.

COROLLARY. If spherical $\triangle A'B'C'$ is the polar triangle of spherical $\triangle ABC$, then $\angle A + \overset{\frown}{B'C'} = 180°$.

Proof. As in Fig. 14.9, extend $\overset{\frown}{AB}$ and $\overset{\frown}{AC}$ along their respective great circles to intersect $\overset{\frown}{B'C'}$ at D and E, respectively. From the theorem, B' is a pole of the great circle of $\overset{\frown}{ACE}$ and thus, by the introductory lemma of this section, $\overset{\frown}{B'E} = 90°$. Similarly, C' is a pole of the great circle of $\overset{\frown}{ABD}$ and $\overset{\frown}{C'D} = 90°$.

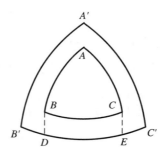

Figure 14.9: $\angle A + \overset{\frown}{B'C'} = 180°$

By addition, $\overset{\frown}{B'E} + \overset{\frown}{C'D} = 180°$. But $\overset{\frown}{B'E} = \overset{\frown}{B'D} + \overset{\frown}{DE}$, so $\overset{\frown}{B'E} + \overset{\frown}{C'D} = (\overset{\frown}{B'D} + \overset{\frown}{DE}) + \overset{\frown}{C'D} = (\overset{\frown}{B'D} + \overset{\frown}{C'D}) + \overset{\frown}{DE} = \overset{\frown}{B'C'} + \overset{\frown}{DE} = 180°$. However, A is a pole of the great circle of $\overset{\frown}{B'C'}$ or, equivalently, A is a pole of the great circle of $\overset{\frown}{DE}$; consequently, from a theorem in Section 14.2, $\angle A = \overset{\frown}{DE}$. If we substitute this into the last relation we obtain $\overset{\frown}{B'C'} + \angle A = 180°$, which we wanted to show.   □

We can now demonstrate the rather surprising result that is the main theorem of this section: The angle sum of a spherical triangle always exceeds 180° and never exceeds 540°.

THEOREM. If $\triangle ABC$ is a spherical triangle, then

$$180° < \angle A + \angle B + \angle C < 540°.$$

Proof. Let $\triangle A'B'C'$ be the polar triangle of $\triangle ABC$. From the last corollary,

$$\angle A + \overset{\frown}{B'C'} = 180°,$$

$$\angle B + \overset{\frown}{A'C'} = 180°,$$

and

$$\angle C + \overset{\frown}{A'B'} = 180°.$$

Add these three equations to produce

$$\angle A + \angle B + \angle C + \overset{\frown}{B'C'} + \overset{\frown}{A'C'} + \overset{\frown}{A'B'} = 540°. \qquad (14.1)$$

However, from the previous section, we know the sum of the sides of a spherical triangle is less than 360°: $\overset{\frown}{B'C'} + \overset{\frown}{A'C'} + \overset{\frown}{A'B'} < 360°$. If we subtract the relation 14.1 from this inequality, we conclude that $\angle A + \angle B + \angle C > 180°$, which is half of what we want. For the other half, just note that $\overset{\frown}{B'C'} + \overset{\frown}{A'C'} + \overset{\frown}{A'B'} > 0°$ in 14.1 and we immediately get $\angle A + \angle B + \angle C < 540°$.   □

14.4   CONGRUENCE THEOREMS FOR TRIANGLES

It will probably come as no surprise that analogues of the familiar SAS, ASA, and SSS triangle congruence theorems from plane geometry also are true on the sphere. It may come as a mild shock, though, that there is an AAA congruence theorem for spherical triangles. This means, in particular, that there is no similarity theory.

Our first step toward discussing these matters is to define two spherical triangles, $\triangle ABC$ and $\angle A'B'C'$, to be congruent if the following six conditions hold when the corresponding sides and angles are arranged in the same order (Fig. 14.10 (a) and (b)).

(1) $\angle A = \angle A'$

(2) $\angle B = \angle B'$

(3) $\angle C = \angle C'$

(4) $\overset{\frown}{AB} = \overset{\frown}{A'B'}$

(5) $\overset{\frown}{AC} = \overset{\frown}{A'C'}$

(6) $\overset{\frown}{BC} = \overset{\frown}{B'C'}$

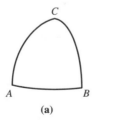 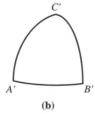 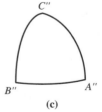

(a)	(b)	(c)

Figure 14.10: Congruent and Symmetric Triangles

The fundamental idea behind congruence is that $\triangle A'B'C'$ can be moved to coincide with $\triangle ABC$ without any deformation of the curvature. Notice, however, that if we form $\triangle A''B''C''$, which is a mirror image of $\triangle ABC$ [see Figures 14.10 (a) and (c)], then the vertices A'', B'', and C'' occur in a clockwise ordering when the ordering of A, B, and C is counterclockwise. This leads to the definition of symmetric spherical triangles.

Define spherical triangles $\triangle ABC$ and $\triangle A''B''C''$ to be *symmetric* if the sides and angles of the first are equal respectively to the sides and angles of the second but are arranged in the reverse order. Namely, the following six conditions hold in the opposite ordering:

(1) $\angle A = \angle A''$

(2) $\angle B = \angle B''$

(3) $\angle C = \angle C''$

(4) $\overset{\frown}{AB} = \overset{\frown}{A''B''}$

(5) $\overset{\frown}{AC} = \overset{\frown}{A''C''}$

(6) $\overset{\frown}{BC} = \overset{\frown}{B''C''}$

The idea here is that $\triangle A''B''C''$ cannot simply be "flipped over" to coincide with $\triangle ABC$ without surface deformation. This problem does not arise when taking mirror images of plane triangles because the plane, in contrast to the sphere, has no curvature. This observation gets to the heart of objections made by mathematicians concerning Euclid's proof of SAS. When a triangle is moved from one place to another in flat space its surface curvature does not change; but a spherical triangle, shifted to a different position on a sphere, may undergo some change and Euclid does not take this into account in his axioms. For example, consider Fig. 14.11.

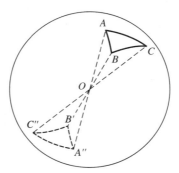

Figure 14.11: Symmetric triangles on a sphere

Suppose the two triangles $\triangle ABC$ and $\triangle A''B''C''$ are situated on the sphere such that $\overline{AA''}$, $\overline{BB''}$, and $\overline{CC''}$ are each diameters. Then it is possible to show these two triangles are symmetric.

The SSS result follows almost free of charge from a previous trihedral result in Chapter 12.

THEOREM. If two triangles on a sphere have the three sides of the first equal respectively to the three sides of the other, then the triangles are either congruent or symmetric. The triangles are congruent if corresponding sides have the same consecutive ordering and are symmetric if they have the reverse ordering.

Proof. Given $\triangle ABC$ and $\triangle A'B'C'$ with $\overgroup{AB}=\overgroup{A'B'}$, $\overgroup{AC}=\overgroup{A'C'}$ and $\overgroup{BC}=\overgroup{B'C'}$, connect the center of the sphere to the respective vertices by constructing $\overline{OA}$, $\overline{OB}$, $\overline{OC}$, $\overline{OA'}$, $\overline{OB'}$, and $\overline{OC'}$ (Fig. 14.12).

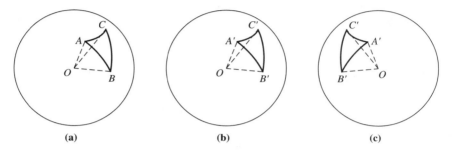

Figure 14.12: Proof of SSS

By the assumed equality of the sides we have $\angle AOB = \angle A'O'B'$, $\angle AOC = \angle A'O'C'$, and $\angle BOC = \angle B'O'C'$. Since the face angles are equal, the trihedral angle $\angle OABC$ is either congruent to $\angle OA'B'C'$ or symmetric to $\angle OA'C'B'$ (see the last theorem of Chapter 12). This means that the corresponding spherical triangles are either congruent [Figures 14.12 (a),(b)] or symmetric [Figures 14.12 (a),(c)] as claimed. □

With SSS in hand, we can give a very slick proof of AAA.

THEOREM. If $\triangle ABC$ and $\triangle DEF$ are triangles on a sphere with $\angle A = \angle D$, $\angle B = \angle E$, and $\angle C = \angle F$, then $\triangle ABC$ is either congruent or symmetric to $\triangle DEF$ depending upon whether the corresponding angles are arranged in the same or opposite order.

Proof. Let $\triangle A'B'C'$ and $\triangle D'E'F'$ be the polar triangles of $\triangle ABC$ and $\triangle DEF$, respectively (Fig. 14.13).

By a previous result, $\angle A$ is the supplement of $\overgroup{B'C'}$ and $\angle D$ is the supplement of $\overgroup{E'F'}$; thus, since $\angle A = \angle D$, we have $\overgroup{B'C'}=\overgroup{E'F'}$. Similarly, we have $\overgroup{A'C'}=\overgroup{D'F'}$ and $\overgroup{A'B'}=\overgroup{D'E'}$. Therefore, by SSS, spherical $\triangle A'B'C'$ is either congruent or symmetric to spherical $\triangle D'E'F'$. Since corresponding parts are equal, we obtain $\angle A' = \angle D'$, $\angle B' = \angle E'$, and $\angle C' = \angle F'$. Now we may use the duality of the polar triangle relation again to see that $\angle A'$ is the supplement

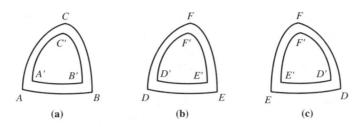

Figure 14.13: Proof of AAA

of $\overset{\frown}{BC}$ and $\angle D'$ is the supplement of $\overset{\frown}{EF}$. Hence $\overset{\frown}{BC}=\overset{\frown}{EF}$ and, in a like manner, $\overset{\frown}{AC}=\overset{\frown}{DF}$ and $\overset{\frown}{AB}=\overset{\frown}{DE}$. Therefore, by SSS again, $\triangle ABC$ is congruent or symmetric to $\triangle DEF$. $\qquad\qquad\square$

Similarly, we may prove the spherical analogues of SAS and ASA.

THEOREM. If $\triangle ABC$ and $\triangle DEF$ are triangles on a sphere with $\overset{\frown}{AC}=\overset{\frown}{DF}$, $\angle A = \angle D$, and $\overset{\frown}{AB}=\overset{\frown}{DE}$, then the two triangles are congruent if these parts are arranged in the same order and symmetric if the parts are arranged in reverse order.

THEOREM. If $\triangle ABC$ and $\triangle DEF$ are triangles on a sphere with $\angle A = \angle D$, $\overset{\frown}{AB}=\overset{\frown}{DE}$, and $\angle B = \angle E$, then the two triangles are congruent if these parts are arranged in the same order and symmetric if these parts are arranged in reverse order.

As consequences of these congruence theorems for spherical triangles, we can easily state corresponding congruence theorems for trihedral angles, based on our study of the previous chapter. For example, the corresponding theorem of AAA for trihedral angles is as follows: Two trihedral angles $\angle AB_1B_2B_3$ and $\angle CD_1D_2D_3$ are congruent if $\angle B_1AB_2 = \angle D_1CD_2$, and $\angle B_2AB_3 = \angle D_2CD_3$, and $\angle B_3AB_1 = \angle D_3CD_1$. We leave it as an exercise for you to state similar corresponding theorems for ASA and SAS.

14.5 AREAS OF SPHERICAL TRIANGLES

The main purpose of this section is to express the area of a spherical triangle in terms of spherical degrees. We assume that our triangles are on a fixed sphere with center O and we will work with degrees as an angle measure, rather than with radians. We need a two-dimensional unit of measure upon which to base our consideration of area. We thus define one *spherical degree* as the area of the spherical $\triangle NPQ$ (Fig. 14.14), where $\overset{\frown}{NP}=\overset{\frown}{NQ}= 90°$ and $\overset{\frown}{PQ}= 1°$. We will postulate that the areas of any congruent or symmetric triangles are equal to each other.

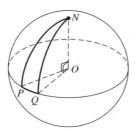

Figure 14.14: Area of $\angle NPQ$ equals one spherical degree

As a small digression we note that, apparently, a spherical degree on one sphere will differ from a spherical degree on a sphere of different radius. If we used the radian as angle measure, then we could define a spherical area unit for a sphere of radius one and deduce that it expands to size r^2 on a sphere of radius r.

With our notion of spherical degrees, it is clear that the spherical $\triangle ABC$ satisfying $\overset{\frown}{AB} = \overset{\frown}{AC} = 90°$ and $\overset{\frown}{BC} = \alpha°$ has an area of α spherical degrees, since areas add for abutting figures. Thus a hemisphere has an area of 360 spherical degrees and a sphere contains 720 spherical degrees.

A *lune* is a figure on a sphere formed by two semicircular arcs of great circles intersecting at antipodal points. Thus, in Fig. 14.15, $ABCD$ is a lune and $\overset{\frown}{ABC}$ and $\overset{\frown}{ADC}$ are semicircles. The angle of a lune is the dihedral angle formed by the half

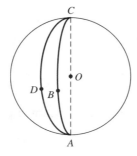

Figure 14.15: Lune $ABCD$

planes containing the semicircles. The proof of the next lemma is immediate from the foregoing remarks about spherical degree.

> **Lemma.** The area of a lune in spherical degrees is twice the number of degrees in the angle of the lune.

We can now state and derive the result we seek. It is neatest to state it in terms of "spherical excess." Recall that the angle sum of a spherical triangle always exceeds 180°. Define the *spherical excess* of spherical $\triangle ABC$ as $\angle A + \angle B + \angle C - 180°$.

THEOREM. The area of a spherical triangle in spherical degrees is numerically equal to the number of degrees in its spherical excess.

Proof. Let $\triangle ABC$ be a spherical triangle and suppose the planes of the great circles containing its sides intersect in diameters $\overline{AD}$, $\overline{BG}$, and $\overline{CF}$, as in Fig. 14.16.

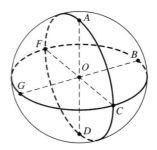

Figure 14.16: Proof of area theorem

From SAS in the plane, $\triangle AOB \cong \triangle GOD$, so $\overline{AB} \cong \overline{GD}$, and since equal chords intercept equal arcs, $\overset{\frown}{AB} = \overset{\frown}{GD}$. LIkewise, $\overset{\frown}{AF} = \overset{\frown}{CD}$ and $\overset{\frown}{BF} = \overset{\frown}{GC}$. By SSS for spherical triangles, spherical $\triangle ABF$ is congruent or symmetric to spherical $\triangle GDC$ and thus, by our postulate, these spherical triangles have equal areas.

Observe that the area of $\triangle ABC$ + the area of $\triangle ABF$ = area of lune $CBFA$ = $2\angle C$ spherical degrees, from the lemma. Since area($\triangle ABF$) = area($\triangle GDC$),

$$\text{area}(\triangle ABC) + \text{area}(\triangle GDC) = 2\angle C \text{ spherical degrees.} \qquad (14.2)$$

Because $\overset{\frown}{ABD}$ and $\overset{\frown}{ACD}$ are semicircles, we have

$$\text{area}(\triangle ABC) + \text{area}(\triangle BCD) = \text{area(lune } ACDB) = 2\angle A \text{ spherical degrees,} \qquad (14.3)$$

and, since $\overset{\frown}{BCG}$ and $\overset{\frown}{BAG}$ are semicircles,

$$\text{area}(\triangle ABC) + \text{area}(\triangle ACG) = \text{area(lune } BAGC) = 2\angle B \text{ spherical degrees.} \qquad (14.4)$$

If we add 14.2, 14.3, and 14.4 the result may be written $2 \cdot \text{area}(\triangle ABC) +$ [area($\triangle ABC$)+area($\triangle GDC$)+area($\triangle BCD$)+area($\triangle ACG$)] = $2(\angle A + \angle B + \angle C)$ spherical degrees. However, the bracketed term is the area of a hemisphere or $360°$ spherical degrees. Hence, $2 \cdot \text{area}(\triangle ABC) + 360$ spherical degrees $= 2(\angle A + \angle B + \angle C)$ spherical degrees, or area($\triangle ABC$) $= \angle A + \angle B + \angle C - 180$ spherical degrees = the spherical excess of $\triangle ABC$, as asserted. $\square$

14.6   A NON-EUCLIDEAN MODEL

We have suggested in previous sections that it might be possible to define a type of "geometry" on the surface of a sphere by interpreting points on the surface as "points" in Euclid's axioms and great circles on the sphere as "lines." The most striking consequence of this attempt is that the Euclidean parallel postulate completely fails! We cannot construct a line through a point that is parallel to a given line because there are not any parallels: Every two lines (great circles) intersect.

Observe that a line segment in this "geometry" may be produced continuously (Euclid's second postulate). Although the great circles are finite in length and not infinite (as are lines in the plane), they are boundaryless, or without end. Euclid's first postulate, which is usually taken to state that two points determine a unique line, also fails in this interpretation, since two antipodal points do not determine a unique great circle. It might occur to you that if we modified our interpretations of points and lines in such a manner that Euclid's (P1) were true, then the parallel postulate (P5) would also be true. In fact, we can alter our model in such a way as to satisfy (P1), but still not (P5). The new "plane" we introduce is an instance of a "Riemannian plane," named in honor of the mathematician B. Riemann, who introduced such ideas in a famous lecture in 1854.

A point of our new model is defined to be a pair of antipodal points of the sphere. Specifically, if A and A' are antipodal points of a sphere, then let $\bar{A} = \{A, A'\}$. The points of our new plane are all such pairs $\bar{A}$. A line in our new plane is the set consisting of all antipodal point pairs that lie on the same great circle. That is, if C is a great circle on the sphere, then $\bar{C} = \{\bar{A} | A \text{ is on } C\}$ is a line in our new plane. We can show that in our new plane, unlike great circles on the sphere, two distinct new lines intersect in one point. Moreover, two antipodal point pairs determine a unique plane through the center of the sphere; thus, a unique line is determined in our new plane and (P1) is satisfied. Unfortunately, our observation about distinct lines meeting means that (P5) also fails in this new model.

PROBLEMS

14.1. Show that a line cannot intersect a sphere in three points.

14.2. The radius of a sphere is 20 units. Find the area of a circle formed by a plane passing through the sphere six units from the center.

14.3. Find the spherical distance between two points on a sphere whose radius is 10 units if the chord joining the points is 10 units.

14.4. A plane and a sphere are tangent to each other if they have one and only one point in common. Prove that if a plane is perpendicular to a radius at its end point on the sphere, then it is tangent to the sphere.

14.5. Prove that the vertical angles formed by two intersecting great circles are equal.

14.6. A great circle of a sphere passes through one end point of a diameter of the sphere. Show that it also passes through the other end point.

14.7. Suppose spherical $\triangle ABC$ on a sphere with center O has associated trihedral $\angle OABC$ and $\angle AOC = \angle BOC$. Prove that $\triangle ABC$ is isosceles.

14.8. If the polar triangle of spherical $\triangle ABC$ coincides with $\triangle ABC$, then find the sum of the angles of $\triangle ABC$.

14.9. Prove that any side of a spherical triangle is a minor arc of a great circle.

14.10. Draw the polar triangle of a spherical triangle whose sides are 90°, 90°, and 60°.

14.11. If two angles of a spherical triangle are equal, then show that its polar triangle is isosceles.

*14.12. Prove that a sphere can be circumscribed about a tetrahedron.

14.13. Which of the following statements are true on the sphere?
  (a) The Hypotenuse-leg theorem
  (b) The angle bisectors of a triangle meet in a point.
  (c) The theorem of Pythagoras
  (d) The area of a right triangle is $\frac{1}{2}bh$.
  (e) Given a line and a point not on the line, there exists a unique perpendicular to the line through the point.
  (f) Given $\triangle ABC$ and $\triangle DEF$ such that $AB = \frac{1}{2}DE$, $BC = \frac{1}{2}EF$, and $AC = \frac{1}{2}DF$, then $\angle A = \angle D$, $\angle B = \angle E$, and $\angle C = \angle F$.

CHAPTER SUMMARY

- Spherical geometry is a geometry in which there are no parallels to a given line.

- A sphere is uniquely determined as the set of points in space at a given distance (the radius length) from a fixed point, called the center of the sphere. Any segment from the center to a point on the sphere is called a radius. A segment through the center joining two points on the sphere is a diameter.

- If a plane π intersects a sphere with center O and radius r a distance $x = \text{dist}(O, \pi)$ from the center, where $0 \le x < r$, then the intersection is a circle with radius $\sqrt{r^2 - x^2}$.

- If $x = 0$, then the plane passes through the center of the sphere and the resulting circle is called a great circle, and the lengths of any radius and diameter of the great circle are the same as that of the sphere. The end points of a diameter of a great circle are called antipodes. Any other circle formed by intersecting a plane and a sphere is called a small circle. Given any circle of the sphere, the diameter of the sphere perpendicular to the plane of this circle passes through its center and is called the axis of the circle. The end points of the axis of any circle are called its poles.

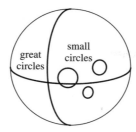

- A unique great circle is determined by any two points of a sphere that are not antipodes.

- Three points on a sphere determine a unique circle of the sphere.

- A spherical angle is an angle formed on the sphere by two intersecting arcs of great circles. The planes of the two great circles containing these arcs each pass through the center of the sphere and intersect in a line containing the common diameter of the circles. In the following figure the arcs $\overset{\frown}{PA}$ and $\overset{\frown}{PB}$ (A and B not antipodes) intersect at P and determine circles and unique planes whose intersection contains the diameter $\overline{PQ}$. The two half planes $\pi_{\text{half}}(A, \overset{\longleftrightarrow}{PQ})$ and $\pi_{\text{half}}(B, \overset{\longleftrightarrow}{PQ})$ determine a dihedral angle with edge $\overset{\longleftrightarrow}{PQ}$. The measure of the spherical angle is the measure of this dihedral angle.

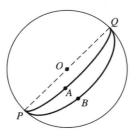

- Let $\angle APB$ be a spherical angle and suppose C is the unique great circle that has P as one of its poles. Extend the arcs $\overset{\frown}{PA}$ and $\overset{\frown}{PB}$ so as to intersect C at A' and B' respectively. Then $\angle APB = \overset{\frown}{A'B'}$ where $\overset{\frown}{A'B'}$ is measured along C, as shown in the preceding figure.

- A spherical triangle is the closed figure formed on a sphere by three arcs of great circles that intersect at their endpoints. In the following figure is spherical $\triangle ABC$ with sides $\overset{\frown}{AB}$, $\overset{\frown}{BC}$ and $\overset{\frown}{AC}$, spherical angles $\angle ABC$, $\angle BCA$, and $\angle CAB$, and vertices A, B, and C.

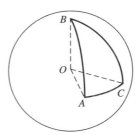

- The spherical $\triangle ABC$ also has an associated trihedral angle formed by connecting the center of the sphere to each of the three vertices by forming radii $\overline{OA}$, $\overline{OB}$, and $\overline{OC}$. The three planes $\pi(O, A, B)$, $\pi(O, A, C)$, and $\pi(O, C, B)$ all intersect at point O to form the trihedral angle $\angle OBAC$. The sides of the spherical triangle, $\overset{\frown}{BA}$, $\overset{\frown}{BC}$, and $\overset{\frown}{AC}$, are measured by the corresponding face angles of the trihedral angle; $\angle BOA$, $\angle BOC$, and $\angle ABC$ are measured by the corresponding dihedral angles of the trihedral angle.

- The sum of any two sides of a spherical triangle is greater than the third

- The sum of the sides of a spherical triangle is less than 360°.

- Two points on a sphere that are not antipodal divide the great circle they determine into the union of two arcs. The shorter of these two arcs is called the minor arc, and it is the shortest polygonal path connecting these points.

- The spherical distance between either pole of a great circle and any point of the great circle is 90°.

- Spherical triangle $\triangle A'B'C'$ is called the polar triangle of spherical $\triangle ABC$ if A is the pole on the A' side of the great circle containing $\overset{\frown}{B'C'}$, B is a pole on the B' side of the great circle of $\overset{\frown}{A'C'}$, and C is a pole on the C' side of the great circle of $\overset{\frown}{A'B'}$, as shown in the following figure. If spherical $\triangle A'B'C'$ is the polar triangle of $\triangle ABC$, then $\triangle ABC$ is the polar triangle of $\triangle A'B'C'$.

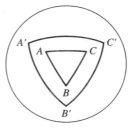

- If spherical $\triangle A'B'C'$ is the polar triangle of spherical $\triangle ABC$, then $\angle A + \overset{\frown}{B'C'} = 180°$. (Each angle of one triangle of a polar triangle pair is the supplement of the opposite side of the other triangle.)

- The angle sum of a spherical triangle always exceeds $180°$ and never exceeds $540°$.

- Two spherical triangles are congruent or symmetric if the following six conditions hold:

  1. $\angle A = \angle A'$,
  2. $\angle B = \angle B'$,
  3. $\angle C = \angle C'$,
  4. $\overset{\frown}{AB} = \overset{\frown}{A'B'}$,
  5. $\overset{\frown}{AC} = \overset{\frown}{A'C'}$,
  6. $\overset{\frown}{BC} = \overset{\frown}{B'C'}$.

- Two spherical triangles are symmetric if the sides and angles of the first are equal respectively to the sides and angles of the second, but are arranged in the reverse order. (They cannot be made to coincide without surface curvature deformation.) They are congruent if the corresponding sides and angles are arranged in the same order.

- If two triangles on a sphere have the three sides of the first equal respectively to the three sides of the other, then the triangles are either congruent or symmetric. The triangles are congruent if corresponding sides have the same consecutive ordering and are symmetric if they have the reverse ordering. (This is SSS for spherical triangles. Similar analogues exist for ASA and SAS.)

- If $\triangle ABC$ and $\triangle DEF$ are triangles on a sphere with $\angle A = \angle D$, $\angle B = \angle E$, and $\angle C = \angle F$, then $\triangle ABC$ is either congruent to or symmetric to $\triangle DEF$ depending upon how the corresponding parts are ordered. (This is the AAA theorem for spherical triangles.)

- When considering areas of spherical triangles, we assume our triangles are on a fixed sphere with center O, and work in degrees. One spherical degree is the area of $\triangle NPQ$ (as shown in the figure on page 210) where $\overset{\frown}{NP} = \overset{\frown}{NQ} = 90°$ and $\overset{\frown}{PQ} = 1°$. A hemisphere has 360 spherical degrees, and a sphere has 720 spherical degrees.

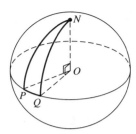

- A lune is a figure on a sphere formed by two semicircular arcs of great circles intersecting at antipodal points. The angle of a lune is the dihedral angle formed by the half planes containing the semicircles, as shown in the following figure.

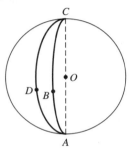

- The area of a lune is twice the number of degrees in the angle of the lune. The spherical excess of spherical $\triangle ABC$ is $\angle A + \angle B + \angle C - 180°$. The area of a spherical triangle in spherical degrees is numerically equal to the number of degrees in its spherical excess.

- In spherical geometry, we have been thinking of great circles as lines. This presents problems with our intuition since great circles have finite length and any two great circles intersect twice, thus removing the possibility of parallels and thereby contradicting Euclid's (P5). If we define a new plane where points are considered to be pairs of antipodal points on the sphere, then any two great circles (lines) intersect in only one point. We have not succeeded in rectifying the failing of (P5) but have instead produced a model of what is known as Riemannian geometry, which is our first non-Euclidean encounter.

CHAPTER 15

Models for Hyperbolic Geometry

15.1 ABSOLUTE GEOMETRY

We alluded to some omissions in Euclid's list of axioms in Chapters 1 and 12. It was not complete enough to furnish a foundation for his later results. In Chapter 14 we have seen that the words "point" and "line" may be given interpretations on the sphere in which some of Euclid's axioms are not true. In particular, we saw that the Euclidean parallel postulate failed on the sphere when great circles were interpreted as lines; in this sense of line there are no parallel lines on the sphere. Several mathematicians since the time of Euclid have extended his axiom list to provide a complete basis for what we know as Euclidean geometry. A famous axiom set that is complete was given by Hilbert around 1899. It turns out that a number of axioms, besides the parallel postulate, fail on the sphere and also on the Riemannian geometry model we derived from the sphere. It would therefore be natural for you to wonder if it is possible to find a model whose point and line interpretations satisfy all the axioms of a complete set, except for the parallel postulate.

Mathematicians have been uncomfortable with the parallel postulate since the time of the ancient Greeks and even Euclid himself did not employ it until after he derived his first 28 propositions. There were a number of attempts over two millenia to establish the parallel postulate (P5) as a consequence of the other Euclidean axioms, but we now know these were doomed to fail. In fact there is a geometry due to Bolyai, Gauss, and Lobachevsky (the latter had the honor of first publication in 1829) that obeys every one of Hilbert's axioms but (P5). In this geometry, (P5) is replaced by assuming that there is a line, ℓ, and a point, P, not on ℓ, such that there are at least two distinct lines through P and parallel to ℓ. This geometry is called hyperbolic geometry, and we will consider two models for it later in this chapter.

In this section we study some results of absolute geometry, which is the body of results derived from the Hilbert axiom set with the parallel axiom omitted (and not replaced by anything else). The results of absolute geometry are valid in both Euclidean geometry and hyperbolic geometry. The main theorems we consider here are due to G. Saccheri and A.-M. Legendre and arose in the course of vain attempts to deduce (P5) as a theorem. Our start along this path is a lemma whose proof follows from the exterior angle theorem given in Chapter 1. We leave its proof as an exercise.

LEMMA. In any triangle in the plane, the sum of any two angles is less than 180°.

The proof of the next lemma goes back to Legendre around 1794 and starts out very much like our proof of the exterior angle theorem.

LEMMA. Suppose $\angle A$ is the (or a) smallest angle of $\triangle ABC$. Then there is a new triangle with the same angle sum as $\triangle ABC$ but with one angle at most one-half of $\angle A$.

Proof. As in Fig. 15.1, let E be the midpoint of $\overline{BC}$ and let F be the unique point on $\overline{AE}$ extended (to the indicated half plane) such that $AE = EF$.

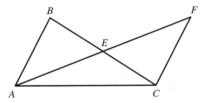

Figure 15.1: Smaller angle but same angle sum

By SAS, we have $\triangle BAE \cong \triangle CFE$ and, from corresponding parts, $\angle B = \angle ECF$ and $\angle BAE = \angle F$.

$$\begin{aligned}
\text{The angle sum of } \triangle ABC &= \angle BAC + \angle B + \angle BCA \\
&= (\angle BAE + \angle FAC) + \angle B + \angle BCA \\
&= \angle F + \angle FAC + (\angle ECF + \angle BCA) \\
&= \angle F + \angle FAC + \angle FCA \\
&= \text{angle sum of } \triangle AFC.
\end{aligned}$$

Moreover, $\angle A = \angle BAE + \angle FAC$, which means that one of $\angle BAE = \angle F$ or $\angle FAC$ is $\leq \frac{1}{2}\angle A$. Thus, $\triangle AFC$ serves as the new triangle we seek, since it has the same angle sum as $\triangle ABC$ and an angle less than or equal to one-half of $\angle A$. This proves the lemma. $\qquad\square$

THEOREM. If $\triangle ABC$ is any triangle, then $\angle A + \angle B + \angle C \leq 180°$.

Proof. Suppose the contrary, that there is some $\triangle ABC$ with $\angle A + \angle B + \angle C = 180° + e°$, where $e > 0$. We can always arrange to reletter the vertices of this triangle so that $\angle A$ is its (or a) smallest angle. But from the lemma immediately preceding, we can replace $\triangle ABC$ with another triangle with angle sum $180° + e°$, but with smallest angle $\leq \frac{1}{2}\angle A$. In turn, we replace this last triangle by yet another new triangle with angle sum $= 180° + e°$ and smallest angle $\leq \frac{1}{2}(\frac{1}{2}\angle A) = \frac{1}{4}\angle A$.

This procedure may be iterated n times to produce a triangle of angle sum $= 180° + e°$ and smallest angle $\leq \frac{1}{2^n} \cdot \angle A$. It follows from a property of the real numbers (the Archimedean property) that we can find some natural number n so that $\frac{1}{2^n} \cdot \angle A < e$. For this value of n we have a triangle whose angle sum is $180° + e°$ and an angle $< e°$. Consequently, the sum of the other two angles of this triangle is greater than $180°$. But this contradicts our initial lemma. Hence, no triangle may have angle sum $> 180°$ and the theorem is proved.   □

This theorem probably strikes you as peculiar, especially since we proved a theorem in Chapter 2 to the effect that the angle sum in any triangle is $180°$. However, in Chapter 2 we assumed (P5), the parallel postulate—here we do not. As we shall see, an angle sum equal to $180°$ is characteristic of Euclidean geometry and an angle sum $< 180°$ is true in hyperbolic geometry.

The concept of defect is useful for our purposes. The defect of an arbitrary $\triangle ABC$ is $180° - (\angle A + \angle B + \angle C)$. It is analogous to the concept of excess introduced in the previous chapter and has a nice connection with area that we will expore in Section 15.4. The proof of the next lemma follows by applying the definition of defect to to both sides of the relation and is left as an exercise.

LEMMA. If D is an interior point of the side $\overline{BC}$ in $\triangle ABC$, then the defect of $\triangle ABC$ equals the sum of the defects of $\triangle ABD$ and $\triangle ADC$.

The next theorem is a startling result of absolute geometry because it shows that if we can measure the angle sum of any one triangle, then we have information about all triangles. The proof is rather long and proceeds in a neat stepwise fashion. The method itself is fascinating and also goes back to Legendre.

THEOREM. If the angle sum of some triangle is $180°$, then the angle sum of every triangle is $180°$.

Proof. Suppose $\triangle ABC$ is a triangle with angle sum $= 180° = \angle A + \angle B + \angle C$. Our first step will be to find a right triangle with angle sum equal to $180°$. If $\triangle ABC$ is a right triangle, then we are done. If $\triangle ABC$ is not a right triangle, then we use the notion of defect as a technical tool to do the job. Denote the defect of any $\triangle UVW$ by $\mathcal{D}(UVW)$; that is,

$$\mathcal{D}(UVW) = 180° - (\angle U + \angle V + \angle W).$$

The previous theorem showed that defect was nonnegative and the previous lemma showed that defect was additive when a triangle was the union of two abutting triangles. If necessary, reletter so that $\angle A$ is the (or a) largest angle of $\triangle ABC$. Since $\triangle ABC$ has angle sum $= 180°$, it is clear that $\mathcal{D}(ABC) = 0$. The altitude from vertex A must intersect $\overline{BC}$ in an interior point E, as in Fig. 15.2. (Why is this true?)

From the additivity of defect, $\mathcal{D}(ABC) = 0 = \mathcal{D}(ABE) + \mathcal{D}(AEC)$, and from the nonnegativity of defect, $\mathcal{D}(ABE) = \mathcal{D}(AEC) = 0$. So we have found a right triangle, say $\triangle ABE$, with angle sum of $180°$ because its defect is zero.

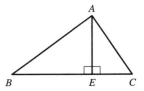

Figure 15.2: $\mathcal{D}(ABC) = \mathcal{D}(ABE) + \mathcal{D}(AEC)$

We can now adjoin to $\triangle ABE$ a congruent right triangle in such a way that their hypotenuses coincide in order to form a rectangle, R, as in Fig. 15.3(a). (Justify this for yourself.) Since right triangles of defect zero were used to construct R, the angle sum of R must be 360°.

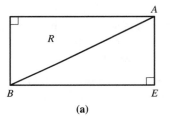

(a)

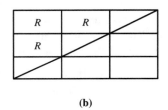

(b)

Figure 15.3: Rectangle R used as a paving tile

Because R is a rectangle, we can use it in paving tile fashion with abutting congruent pieces to form a rectangle of arbitrarily large dimensions, as shown in Fig. 15.3(b). (Technically, this, too, depends upon the Archimedean property of the reals: If we add any length to itself repeatedly, then we can cover an arbitrarily large distance.) Moreover, any of these arbitrarily large rectangles has angle sum = 360° and is divided by a diagonal into two right triangles, each with angle sum = 180° or defect = 0. Thus, we have completed our second step, which consisted of using one right triangle of defect zero to construct an arbitrarily large right triangle of defect zero.

As step 3 we now show that every right triangle has defect zero. Let $\triangle XYZ$ be a right triangle with right angle at vertex Y. Our intent is to show that $\mathcal{D}(XYZ) = 0$. To prove this, observe that our second step allows us to construct a right triangle, $\triangle UYW$, of defect zero, whose legs contain points X and Z as interior points, as in Fig. 15.4. If we construct $\overline{UZ}$ and use what we know about the additivity of defect, we get

$$\mathcal{D}(UYW) = 0$$
$$= \mathcal{D}(UZW) + \mathcal{D}(UYZ) = \mathcal{D}(UZW) + \mathcal{D}(UXZ) + \mathcal{D}(XYZ),$$

from which we conclude that $\mathcal{D}(XYZ) = 0$, as desired.

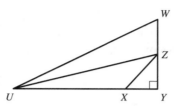

Figure 15.4: Embedding a right triangle in a larger right triangle

The fourth step is just the punch line to complete the proof of the theorem. Every triangle may be sectioned into two right triangles, each of defect zero, by simply dropping an altitude. By additivity of the defect, every triangle thus has defect zero, or angle sum equal 180°. □

We now see that models for absolute geometry may be classified according to the angle sum of their triangles.

COROLLARY. In any model for the axioms of absolute geometry either the angle sum of any triangle is 180° or the angle sum of any triangle is less than 180°, not both.

Proof. Let $\triangle ABC$ be an arbitrary triangle. If $\angle A + \angle B + \angle C = 180°$, then the theorem tells us that any triangle has angle sum $= 180°$. On the other hand, if $\angle A + \angle B + \angle C < 180°$, then no other triangle may have angle sum $\geq 180°$: Angle sums greater than 180° were foreclosed by our first theorem and a sum of 180° would produce a contradiction. □

15.2   THE KLEIN-BELTRAMI DISK

In the last section we mentioned that Hilbert had given a complete axiom set for Euclidean geometry. In his list, Hilbert incorporated as a parallel axiom the Playfair formulation of the (P5) parallel postulate of Euclid (see Exercise 7 of Chapter 2). If we substitute the following hyperbolic axiom for the parallel axiom and leave the rest of Hilbert's list unchanged, then we obtain the body of results known as hyperbolic geometry. The hyperbolic axiom states that there exists a line and a point not on this line through which there pass at least two parallels to this line.

Naturally, the results that are deducible in this altered axiom system are as deep and varied as the propositions of Euclidean geometry. We mention two of these for our purposes here. First, it can be shown in hyperbolic geometry that there exists a triangle whose angle sum is less than 180°. On the basis of the previous theorem, we conclude that every triangle in hyperbolic geometry has angle sum $< 180°$. We already know that the angle sum of any Euclidean geometry triangle is 180°, a result proved using the parallel postulate. Thus, the angle sum of a triangle characterizes the geometry.

The second hyperbolic result we mention is that for every line ℓ and any point P not on ℓ there are at least two distinct lines through P parallel to ℓ. In fact, this may be pushed further to show that there are an infinite number of lines through P that do not meet ℓ. In this section we will define interpretations of the words "point" and "line" in the axioms to form a model introduced toward the end of the nineteenth century and named to honor E. Beltrami and F. Klein. Using this model, we will find it easy to exhibit an illustration of the hyperbolic axiom.

Let C_1 be a circle with fixed center O and radius $r > 0$ in the Euclidean plane. The "points" of the Klein-Beltrami disk model for the hyperbolic plane are the usual points of the Euclidean plane that are interior to C_1. More precisely, the "points" of this model are Euclidean points A such that $OA < r$. If A and B are two points of C_1, then they determine a chord $\overline{AB}$ of the circle. The "lines" of this model will be all open chords of C_1; that is, "lines" are chords $\overline{AB}$ without the end points A and B (Fig. 15.5). Notice that this open chord is the intersection of the Euclidean line $\overleftrightarrow{AB}$ with the interior of C_1. Thus, two distinct open chords, or "lines," intersect in at most one "point" and two "points" determine a unique "line."

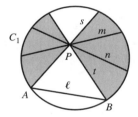

Figure 15.5: The Klein-Beltrami disk model of the hyperbolic plane

With P a "point" not on the open chord ℓ, it is apparent from Fig. 15.5 that open chords m and n are "lines" through P that are parallel to ℓ in the sense that they have no "points" in common with ℓ. The fact that m and n, if extended, might intersect ℓ extended outside C_1 is irrelevant—only the points interior to C_1 comprise the hyperbolic plane. Observe that the open chords s and t determined by the Euclidean lines $\overleftrightarrow{PA}$ and $\overleftrightarrow{PB}$ are also both parallel to ℓ, since points A and B are not "points" of the model. Finally, we remark that any open chord through P which lies in the shaded region determined by s and t is also parallel to ℓ, which shows that infinitely many parallels to ℓ may be passed through P. It is now clear that s, m, n, and t are parallel to ℓ not because of their appearance in a drawing but because they fit the definition of parallel lines.

Together with a suitable interpretation of the "angle" between two "lines" it can be shown that all the axioms of Euclidean geometry, indeed all of Hilbert's axioms, except for (P5) are true in this model. We have seen that the hyperbolic axiom is true here, and an important feature of the model is that it proves that (P5) cannot be deduced as a theorem from the other axioms of Euclid (or Hilbert). For if it were implied by the other axioms and results, it would be a true statement for the model, which is an inherent impossibility. Consequently, the efforts expended to derive (P5) from other axioms, from ancient times through the middle ages and up to the beginning of the nineteenth century, were destined to fail.

15.3   THE POINCARÉ DISK

The model of the previous section provided an easily understood illustration of the hyperbolic axiom; however, the method for measuring angles between "lines" is different from the usual Euclidean one. Thus, we cannot use the Klein-Beltrami model to provide a satisfying example of a triangle with angle sum less than 180°. We will use another disk model for this purpose due to Henri Poincaré (1854–1912) in which measurement of angles is done in the usual Euclidean way. As an introductory remark we point out that an angle made by two intersecting circular arcs is measured by the number of degrees in the angle formed by their tangent lines. Thus, in Fig. 15.6(a) the angle between the two arcs meeting at A is given by $\angle BAC$, where $\overleftrightarrow{BA}$ and $\overleftrightarrow{CA}$ are tangents to the respective circles at A. Similarly, as in Fig. 15.6(b), the angle between

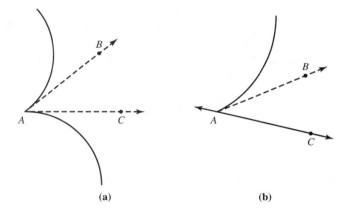

Figure 15.6: Angles between two arcs or an arc and a line

the arc and a line $\overleftrightarrow{CA}$ intersecting at A is $\angle BAC$, where $\overleftrightarrow{BA}$ is tangent to the circle at A.

The "points" of the Poincaré disk are the Euclidean points interior to a fixed circle C_2 with center O, as before. However, here the "lines" arise in two ways. First, "lines" will be open arcs of circles that intersect C_2 at 90° angles. These "lines" are given by the intersection of Euclidean circles orthogonal to C_2 with the interior of C_2. Thus, in Figure 15.7, m is the "line" given by the arc (not including its end points A and C) of the circle intersecting at A and C. Second, "lines" may also be open diameters of C_2; that is, any open chord of C_2 through O is also a "line." This second interpretation of "line" may be viewed as the open arc arising from intersecting a circle of infinite radius with the interior of C_2. Thus, the "line" ℓ in Fig. 15.7 is the open chord $\overline{AB}$ without its end points through O. Moreover, the "lines" m and ℓ are parallel because A is not a "point" of this model of the hyperbolic plane. By interpreting the angle between two "lines" as we indicated in our introductory remark, it can be demonstrated that the

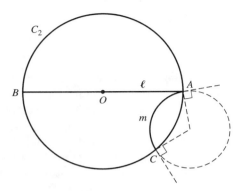

Figure 15.7: The Poincaré disk model of the hyperbolic plane

Poincaré disk satisfies all the axioms defining a hyperbolic plane and is thus a model for hyperbolic geometry. In particular, two "points" of the disk or interior points of C_2 determine a unique open arc, or "line."

We use the last observation to construct a triangle in this Poincaré model (Fig. 15.8). On two perpendicular open diameters, choose "points" A and B of the disk whose Eu-

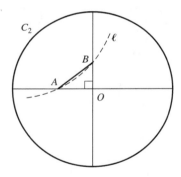

Figure 15.8: A triangle with positive defect

clidean distances from O are equal: $OA = OB$. Now the two "points" A and B determine a unique "line" ℓ of the Poincaré disk. The Euclidean $\triangle BOA$ has angle sum $= 180°$ and the segment of the open arc ℓ that joins B and A lies below the Euclidean segment $\overline{BA}$. The angle that $\overline{OA}$ makes with ℓ is thus less than $45°$, since the tangent to ℓ at A lies between $\overline{AB}$ and $\overline{AO}$, whereas the Euclidean $\angle BAO = 45°$. Consequently, the angle sum of the triangle formed by the three intersecting hyperbolic "lines" is less than $180°$.

The two models we have given of the hyperbolic plane are not the only ones that exist. It is a fact that a 1-1 correspondence may be set up between any two of them that associates "points" and "lines" in each so as to preserve all geometric relations in the models. This means that, in essence, all models of hyperbolic geometry are the same and that the Hilbert axiom system with (P5) replaced by the hyperbolic axiom is what is known as categorical.

15.4 THE AAA THEOREM IN HYPERBOLIC GEOMETRY

There are a number of parallels between hyperbolic geometry and spherical geometry. One that we hinted at earlier is that in hyperbolic geometry the area of a triangle may be taken to be the deficit, just like the area of a triangle in spherical geometry is the excess. There are also striking parallels between the trigonometric formulas in spherical geometry and hyperbolic geometry, although developing these formulas would involve a large detour. A surprising result is that the AAA theorem is also a theorem in hyperbolic geometry.

> **THEOREM(AAA FOR HYPERBOLIC GEOMETRY).** Given $\triangle ABC$ and $\triangle DEF$, each with angle sum less than 180°, suppose that $\angle A \cong \angle D$, $\angle B \cong \angle E$, and $\angle C \cong \angle F$. Then $\triangle ABC \cong \triangle DEF$.

Proof. Observe that if one side of $\triangle ABC$ is congruent to a corresponding side of $\triangle DEF$, then we are done by ASA. Consequently, we face no loss of generality by relettering if necessary to assume $AB > DE$. Moreover, in addition, we may have either $AC > DF$ or $AC < DF$. We will prove the theorem for the case $AB > DE$ and $AC > DF$ and leave the remaining case $AB > DE$ and $AC < DF$ for an exercise.

Since $AB > DE$ and $AC > DF$, we may find interior points G on $\overline{AB}$ and H on $\overline{AC}$, as in Fig. 15.9, such that $AG = DE$ and $AH = DF$.

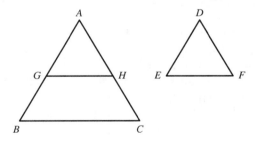

Figure 15.9: Proof of AAA

We use SAS to see that $\triangle AGH \cong \triangle DEF$. Since $\angle AGH = \angle E = \angle B$ and $\angle AHG = \angle F = \angle C$, we must have

$$\angle HGB + \angle B = 180°$$

and

$$\angle GHC + \angle C = 180°.$$

Thus, $GHCB$ is a convex quadrilateral with angle sum $360°$, which contradicts Exercise 4 of this chapter. Consequently, we must have a corresponding pair of equal sides, and AAA is proved. □

15.5 GEOMETRY AND THE PHYSICAL UNIVERSE

You may now feel that hyperbolic geometry is an amusing intellectual exercise but that Euclidean geometry is the one that describes the "real world." Poincaré thought that the question of which geometry was the "true" one was meaningless. He took axioms for each of the geometries to be conventions adopted to model an approximately determined reality and considered only whether one geometry was more convenient than another. (see H. Poincaré, *Science and Hypothesis,* Dover, 1952).

All measurements involve both physical and geometric assumptions. If we measure the angle sum of a huge triangle as Gauss allegedly did, using three mountain peaks as vertices, and detect a defect, then we could claim that space is hyperbolic. On the other hand, we could also claim space to be Euclidean and that light rays do not always travel in straight lines. To maintain his argument, Poincaré devised an imaginary universe, Σ, occupying the interior of a sphere of radius R in Euclidean space in which the following physical laws hold:

1. At any point P of Σ, the absolute temperature T is given by $T = T(P) = k(R^2 - r^2)$, where r is the Euclidean distance from P to the center of Σ and k is a constant.

2. The linear dimensions of a material body vary directly with the absolute temperature of the body's locality.

3. All material bodies immediately assume the temperature of their localities.

It follows from (1) and (2) that an inhabitant of Σ shrinks as he or she approaches the boundary and would thus need an infinite number of steps to reach the boundary. Since the inhabitant could not reach the boundary of Σ after taking any large finite number of steps, the universe would appear to be infinite. Moreover, this shrinkage could not be detected because any measuring stick would also shrink. From (3), the inhabitant would not feel any temperature changes.

Poincaré showed that the paths of shortest length joining two points of Σ were circular arcs orthogonally meeting the boundary of Σ. If it is further assumed that light rays travel along these shortest length paths, then these paths look "straight" to an inhabitant. Now we know from our consideration of the Poincaré disk that the hyperbolic axiom holds in Σ and that therefore the inhabitants would believe they live in a non-Euclidean world. However, we know their world is a piece of Euclidean space. Both viewpoints are essentially correct and the question of which is "true" is meaningless.

According to Einstein, the geometry of space-time is affected by matter, and light rays are curved by the gravitational attraction of masses. Neither Euclidean nor hyperbolic geometry is appropriate to describe the geometry of general relativity theory.

In those portions of space that have negative curvature ("saddle shape") the geometry must be hyperbolic.

PROBLEMS

*15.1. Prove that in any plane triangle the sum of any two angles is less than $180°$.

15.2. The defect of a triangle is the difference between $180°$ and the angle sum of the triangle. Denote the defect of $\triangle ABC$ by $\mathcal{D}(ABC) = 180° - (\angle A + \angle B + \angle C)$. If E is an interior point of side $\overline{BC}$ of $\triangle ABC$, then show that $\mathcal{D}(ABC) = \mathcal{D}(ABE) + \mathcal{D}(AEC)$.

15.3. Let $\angle A$ be the largest angle of $\triangle ABC$. Prove that if E is the foot of the altitude from A to $\overleftrightarrow{BC}$, then E is an interior point of $\overline{BC}$.

15.4. Show that all convex quadrilaterals in hyperbolic geometry have angle sum less than $360°$. (Assume results quoted in the text.)

15.5. Let the midpoints of the sides of $\triangle ABC$ be A', B', and C'. Prove that if the four triangles $\triangle AB'C'$, $\triangle C'A'B$, $\triangle B'CA'$, and $\triangle A'C'B'$ are all congruent, then Euclid's Parallel Postulate holds. What if you only assume that two of them are congruent?

15.6. In the Euclidean plane any three parallel lines have a common transversal. Find three parallel "lines" in the Klein-Beltrami disk model of the hyperbolic plane that do not have a common transversal.

*15.7. Let the Euclidean line $\overleftrightarrow{PQ}$ determined by the two "points" P and Q intersect C_1 in the Klein-Beltrami disk in points S and T in the order S, P, Q, T. Define the hyperbolic distance between P and Q by

$$d(P, Q) = \log \frac{QS \cdot PT}{PS \cdot QT}.$$

(a) Suppose C_1 is the circle given by $x^2 + y^2 = 1$ of center $(0, 0)$ and radius $= 1$ in the usual x, y plane. If $P = (0, 0)$ and $Q = (\frac{1}{2}, 0)$, then find $d(P, Q)$.

(b) Find the coordinates of the "point" M, which is the midpoint in this model of the "line segment" $\overline{PQ}$.

(c) If P, Q, and R are the three "points" on a "line," show that $d(P, R) = d(P, Q) + d(Q, R)$.

(d) Fix P and let "point" Q move along the "line" toward T. Show that $d(P, Q) \longrightarrow \infty$.

15.8. (a) Make a sketch showing that the Poincaré disk model satisfies the hyperbolic axiom for parallelism.

(b) Sketch an example of a triangle in the Poincaré disk whose sides are all segments of open arcs and provide a plausible argument that its angle sum is $< 180°$.

15.9. In each of the following models, state whether the Euclidean parallel postulate holds, the hyperbolic axiom holds, or neither holds and justify your answer.

(a) "Point" means a point on the surface of an open hemisphere (boundary excluded) and "line" means an open great semicircle on this hemisphere.

(b) Let P be a fixed point on the surface of sphere S. "Point" means a point of S different from P and "line" means a circle on S that passes through P but with P deleted.

(c) Let P be a fixed point of usual three-dimensional space. "Point" means a straight line through P and "line" means a plane through P.

(d) Let C be the cone in space pictured in Fig. 15.10.

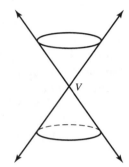

Figure 15.10: Problem 8(d)

"Point" means any line in space that passes through the point V and is otherwise in the interior of C and "line" is the set of all "points" that lie on a plane through V that passes through the interior of C.

15.10. Prove the AAA theorem in the case when $AB > DE$ and $AC < DF$.

CHAPTER SUMMARY

- We know from several previous comments that Euclid's axiom list is incomplete. Moreover, the parallel postulate is independent in the sense that it cannot be derived from any of Euclid's other postulates. However, geometries have been developed by Lobachevsky and others that obey all Euclid's axioms except (P5). Hyperbolic geometry is a geometry in which we assume for some line ℓ and a point P not on ℓ that there are at least two distinct lines parallel to ℓ through P. In absolute geometry, all of Euclid's axioms except (P5) are obeyed. Both types of geometry yield results which seem very different from the Euclidean geometry we've studied so far.

- Absolute geometry results are as follows:

  - In any triangle in the plane, the sum of any two angles is $< 180°$.

– Suppose $\angle A$ is the (or a) smallest angle of $\triangle ABC$. Then there is a new triangle with the same angle sum as $\triangle ABC$ but with one angle at most one-half of $\angle A$.

– If $\triangle ABC$ is any triangle, then $\angle A + \angle B + \angle C \leq 180°$.

– Define the defect of $\triangle ABC$, to be $180° - (\angle A + \angle B + \angle C)$. If D is an interior point of the side $\overline{BC}$ in $\triangle ABC$, then the defect of $\triangle ABC$ equals the sum of the defects of $\triangle ABD$ and $\triangle ADC$.

– If the angle sum of some triangle is $180°$, then the angle sum of every triangle is $180°$. Thus, in any model for the axioms of absolute geometry, either the angle sum of any triangle is $180°$, or the angle sum of every triangle is less than $180°$, but not both.

• The hyperbolic axiom states that there exists a line and a point without the line through which there pass at least two parallels to the line. If we substitute this for (P5) we find that every triangle in hyperbolic geometry has angle sum $< 180°$.

• In the Klein-Beltrami and Poincaré models of hyperbolic geometry, we defined points and lines in such a way that we were able to satisfy all of Euclid's and Hilbert's axioms with the hyperbolic axiom replacing (P5), and we also indicated that (P5) cannot be derived from the other axioms in these models.

• We used the Poincaré disk model of hyperbolic geometry to construct a triangle with angle sum less than $180°$.

• The two models just mentioned are not the only models for hyperbolic geometry. Moreover, all models for hyperbolic geometry are essentially equivalent and the Hilbert axiom system with (P5) replaced by the hyperbolic axiom therefore characterizes hyperbolic geometry (i.e., it is "categorical").

• No one "type" of geometry is any more valid than the other. Each geometry has its place and its uses. Each geometry, then, is "true" to a certain extent. On the sphere, we deal with spherical geometry; in four-dimensional space-time relationships, we would use a geometry that is compatible with Einstein's relativity theory, etc.

- The following figure shows triangles in three different geometries we've considered.

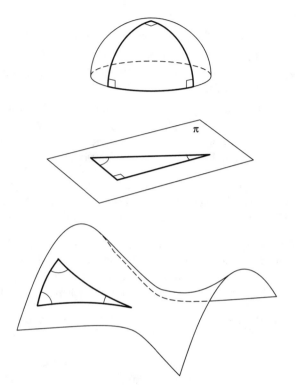

Hints to Starred Exercises

CHAPTER 1

2. Work with the case where G is on the opposite side of $\overleftrightarrow{BC}$ from A and $\overline{AG}$ intersects $\overleftrightarrow{BC}$ at point H, which is strictly between B and C. Show that $\triangle BGC$ is a congruent copy of $\triangle EDF$ and use corresponding parts and transitivity of congruence to get the result.

4. Suppose $\overline{AB}$ and $\overline{DE}$ are not congruent with $AB > DE$. With G in the interior of $\overline{BA}$ so that $\overline{BG} \cong \overline{ED}$, note that $\triangle GBC$ is a congruent copy of $\triangle DEF$. Use the exterior angle theorem to obtain a contradiction.

8. Construct G interior to $\angle BAC$ such that $\angle BAG \cong \angle D$ and $\overline{AG} \cong \overline{DF}$. Work here with the case when G is also interior to $\triangle ABC$. Let the angle bisector of $\angle CAG$ intersect $\overline{BC}$ at H. Show that $\triangle AGH \cong \triangle ACH$, so $HG = HC$. Finally, use the triangle inequality in $\triangle BGH$ to obtain the conclusion.

CHAPTER 2

3. (a) implies (b) is immediate from the text. To show (b) implies (c), just construct diagonal $\overline{DB}$ and make use of the resulting congruent triangles. Finally, to prove (c) implies (a), use the fact that the angle sum of a quadrilateral is $360°$ together with equal angle pairs to conclude that angles A and B sum to $180°$. Thus lines containing opposite sides cannot meet.

4. If $ABCD$ is a rectangle, then opposite sides are of equal length. Since all angles are right angles we can now obtain the result by using congruent triangles. If $\overline{AC} \cong \overline{BD}$, then use SSS to get congruent triangles and consequently equal adjacent angles which sum to $180°$.

CHAPTER 3

5. Write two expressions for the area of the triangle.

6. Think Pythagoras!

12. Work with the case when angle D is obtuse. Let G be the foot of the perpendicular from F to $\overleftrightarrow{DE}$. Use results in the text to relate ratios of the areas of $\triangle ABC$, $\triangle DFG$, and $\triangle DEF$.

CHAPTER 4

2. Construct $\triangle D'E'F'$ with $\angle E' \cong \angle B$, $D'E' = DE$, and $E'F' = EF$ and use the technique of the text proofs.

3. Consider case (b) of Figure 4.8 and show that (a) implies (b), (b) implies (c), (c) implies (d), and (d) implies (a). To show (c) implies (d), invert both sides of the given equation to get

$$\frac{EC}{EA} = \frac{EG}{ED}, \text{ consequently } \frac{EA - AC}{EA} = \frac{EB - BD}{EB},$$

which may be manipulated to produce (d).

9. Let B be any point in S. Let ℓ' be the line through B and parallel to ℓ. Prove that $S = \ell'$ using exercise 3.

CHAPTER 5

3. Use the fact that tangent segments to a circle from an external point are congruent.

4. Let O' be the center of the smaller triangle and compare $\triangle ABO'$ to $\triangle ACO$.

7. Find two expressions for the square of the length of the tangent line.

10. If X is another point on the P side of the circle, then use the theorem that an angle inscribed in a semicircle is a right angle to conclude that $\angle PQX$ is always obtuse and $\angle PXQ$ is always acute.

11. One way to do the exercise would be to use the hint of exercise 7.

20. Use lemmas 1 and 2.

CHAPTER 6

2. From the fact that $2 \cdot 3 - 1 \cdot 5 = 1$ conclude that $2(\frac{1}{5}) - 1(\frac{1}{3}) = \frac{1}{15}$. Now apply this to the fact that $60°$ and $36°$ angles are constructible.

4. There are 12.

CHAPTER 7

5. First prove the "only if " part. To prove the "if " part, let (O) be the circumcircle of $\triangle ABC$. If D is not on (O), get a contradiction.

9. Two tangents to a circle from a point are congruent.

12. Use Euler's theorem.

CHAPTER 8

5. Extend $\overline{AA'}$ to a point E with $\overline{AA'} \cong \overline{EA'}$. What can you say about $ABEC$?

6d. Prove that if $\triangle ABC \sim \triangle DEF$ then the ratio must be $\sqrt{\frac{3}{4}}$.

11. Draw $\overline{AC}$ and $\overline{BD}$.

CHAPTER 9

2. $\overline{AH}$ will be the diameter.

3(b). Use part a.

3(c). Use and generalize part b.

6. Reflect A in ℓ, as in Fagnano's problem.

CHAPTER 10

1. Use the geometric law of cosines.

3. Tangents to a circle from a given point are congruent.

5a. $\angle A = \angle B'AC' = \angle PAC' + \angle PAB'$. Show that $\angle PAC'$ and $\angle PAB'$ are each at least $60°$.

5b. $\angle B = 360° - \angle PBA - \angle PBC$. Show that $\angle PBA$ and $\angle PBC$ are at most $120°$.

11. Reflect A in ℓ.

CHAPTER 11

12. First construct a triangle similar to $\triangle ABC$.

13. Use the information in the order given.

14. See Section 8.2.

15. Use Exercise 8.3(b).

CHAPTER 12

6. See exercise 4.

7. See exercises 4 and 6.

9. See exercise 8.

14. Compare $\overline{EF}$ and $\overline{GH}$ to $\overline{AC}$.

18. The line must lie in $\pi(\ell, A)$ and $\pi(\ell', A)$.)

CHAPTER 14

12. If the tetrahedron is $ABCD$, then let E be the center of the circumcircle of $\triangle ABC$ and let F be th circumcenter of $\triangle BCD$. Show that the perpendiculars to $\pi(A, B, C)$ at E and $\pi(B, C, D)$ at F intersect at an interior point O of the tetrahedron and think about the properties of O.

CHAPTER 15

1. Use the exterior angle theorem.

7. Observe that $d(P, Q) = \log\left[\left(\frac{3}{2}\right)(1)/(1)\left(\frac{1}{2}\right)\right] = \log 3$.

Index